Design of Function Circuits with 555 Timer Integrated Circuit

This text discusses sigma-delta-type function circuits, peak detecting function circuits, and peak sampling function circuits in a detailed manner. It further covers all the function circuits designed by using the basic principles of the six building blocks: integrator, the 555 timer integrated circuit, switch, low pass filter, peak detector, and sample and hold circuit. It is a useful reference text for senior undergraduate and graduate students in the fields of electrical engineering and electronics and communication engineering. This book is accompanied by teaching resources, including a solution manual for the instructors.

- Discusses function circuits such as multipliers, dividers, and multiplier cum dividers using the 555 timer.
- Explains how function circuits are developed with a simple integrator and the 555 timer.
- Extends the applications of 555 timers to perform in function circuits.
- Covers important topics such as monostable multivibrator, inverting amplifier, and peak responding divider.
- Presents function circuit conversion such as multiplier to square root and divider to a multiplier.

This comprehensive book covers the design of function circuits with the help of 555 timer integrated circuits in a single volume. It further discusses how derived function circuits are implemented with integrator, comparator, low pass filter, peak detector, and sample and hold circuits.

Design of Function Circuits with 555 Timer Integrated Circuit

K. C. Selvam

CRC Press
Taylor & Francis Group
Boca Raton London New York

CRC Press is an imprint of the
Taylor & Francis Group, an **informa** business

First edition published 2023
by CRC Press
6000 Broken Sound Parkway NW, Suite 300, Boca Raton, FL 33487–2742

and by CRC Press
4 Park Square, Milton Park, Abingdon, Oxon, OX14 4RN

CRC Press is an imprint of Taylor & Francis Group, LLC

Library of Congress Cataloging-in-Publication Data
Names: Selvam, K. C., author.
Title: Design of function circuits with 555 timer integrated circuit / K.C. Selvam.
Other titles: Design of function circuits with five hundred fifty-five timer integrated circuit
Description: First edition. | Boca Raton : CRC Press, [2023] | Includes bibliographical references and index.
Identifiers: LCCN 2022037988 (print) | LCCN 2022037989 (ebook) | ISBN 9781032391700 (hbk) | ISBN 9781032424798 (pbk) | ISBN 9781003362968 (ebk)
Subjects: LCSH: Function generators (Electronic instruments)—Design and construction. | 555 timer IC (Integrated circuits)
Classification: LCC TK7895.F8 S45 2023 (print) | LCC TK7895.F8 (ebook) | DDC 621.3815—dc23/eng/20221107
LC record available at https://lccn.loc.gov/2022037988
LC ebook record available at https://lccn.loc.gov/2022037989

ISBN: 978-1-032-39170-0 (hbk)
ISBN: 978-1-032-42479-8 (pbk)
ISBN: 978-1-003-36296-8 (ebk)

DOI: 10.1201/9781003362968

Typeset in Sabon
by Apex CoVantage, LLC

Dedicated to my loving wife
S. Latha

Contents

Preface

After writing three books and publishing them with CRC Press, Taylor & Francis, I found that the very popular timer IC 555 can be used to perform function circuits. I worked on that, got useful results, and decided to write another book, and this is the result. Earlier, the 555 timer IC was used for timing and control applications, and now it can also use to perform function circuits.

I am highly indebted to my:

- Mentor, Prof. Dr. V.G.K. Murti who taught me about function circuits.
- Philosopher, Prof. Dr. P. Sankaran who taught me measurements and instrumentation.
- Teacher, Prof. Dr. K. Radha Krishna Rao who taught me operational amplifiers.
- Gurunather, Prof. Dr. V. Jagadeesh Kumar who guided me in the proper way of the scientific world.
- Trainer, Dr. M. Kumaravel who trained me to do experiments with op-amps.
- Director, Prof. Dr. Kamakoti who motivated me to do this work.
- Encourager, Prof. Dr. Enakshi Bhattacharya who encouraged me to get this result.
- Leader, Prof. Dr. Devendra Jalihal who kept me in a happy and peaceful official atmosphere.
- Supervisor, Prof. Dr. David Koil Pillai who supervised all my research work at IIT Madras.

I thank Dr. Gauravjeet Singh Reen, Senior Commissioning Editor, Taylor & Francis, CRC Press who has shown keen interest in publishing all my theory and concepts on function circuits. His hard work in making this book possible is commendable.

I also thank my friends Prof. Dr. R. Sarathi, Dr. Balaji Srinivasan, Dr. T. G. Venkatesh, Dr. Bharath Bhikkaji, Dr. Bobey George, Dr. S. Anirudhan, Dr. Aravind, Mrs. T. Padmavathy, Mrs. Sulochana, and Mrs. Karthiayini for their constant encouragement throughout my research work. I thank all other staff, students, and faculty of the Electrical Engineering Department, Indian Institute of Technology–Madras, for their immense help during the experimental setups, manuscript preparation, and proofreading.

Author Biography

Dr. K. C. Selvam was born on April 2, 1968, in Krishnagiri district of Tamil Nadu State, India. He obtained a diploma in electronics and communication engineering from the government polytechnic college, Krishnagiri, Tamil Nadu, India, in 1986. He graduated from the Institution of Electronics and Telecommunication Engineers, New Delhi, in 1994. He obtained an honorary PhD degree from the University of Swahili, Government of Panama, in the year 2020.

He has been conducting research and development work for the past 33 years and has published more than 33 research papers in various national and international journals. He also published the following technical and scientific books in international publishers.

Design of Analog Multipliers Using Operational Amplifiers (CRC Press, Taylor & Francis, New York and London, July 2019), DOI:10.1201/9780429277450, ISBN: 9780429277450.

Multiplier-cum-Divider Circuits; Principles, Design and Applications (CRC Press, Taylor & Francis, New York and London June 2021), DOI:10.1201/9781003168515, ISBN: 9781003168515.

Analog Function Circuits: Fundamentals, Principles, Design and Applications (CRC Press, Taylor & Francis, New York and London, December 2021), ISBN 9781032081601.

Design of Function Circuits with 555 Timer IC (Accepted and to be published in CRC Press, Taylor & Francis).

Principles of Function Circuits (Lambert Academic Publishing, Germany), ISBN-13: 9786200532411.

Analog Dividing Circuits (Lambert Academic Publishing, Germany), SBN-13: 978–6200653987, ISBN-10: 6200653984.

He got Best Paper Award by IETE in 1996 and the Students Journal Award by IETE in 2017. In 2021, he received the Life Time Achievement Award from the Institute of Researchers, Wayanad, Kerala, India. At present he is working as a scientific staffer in the Department of Electrical Engineering, Indian Institute of Technology–Madras, India.

Useful Notations

V_1	First input voltage
V_2	Second input voltage
V_3	Third input voltage
V_O	Output voltage
V_R	Reference voltage/peak value of first saw tooth waveform
V_T	Peak value of first triangular waveform
V_P	Peak value of second triangular wave/saw tooth wave
V_C	Comparator 1 output voltage in the first saw tooth/triangular wave generator
V_M	Comparator 2 output voltage by comparing saw tooth/triangular waves with one input voltage
V_N	Low pass filter input signal
V_{S1}	First generated saw tooth wave
V_{S2}	Second generated saw tooth wave
V_{T1}	First generated triangular wave
V_{T2}	Second generated triangular wave
V_S	Sampling pulse
V_1'	Slightly less than V_1 voltage
V_2'	Slightly less than V_2 voltage

Abbreviations

TDM	Time division multiplier
MTDM	Multiplexing time division multiplier
STDM	Switching time division multiplier
PRM	Peak responding multiplier
MPRM	Multiplexing peak responding multiplier
SPRM	Switching peak responding multiplier
PDM	Peak detecting multiplier
MPDM	Multiplexing peak detecting multiplier
SPDM	Switching peak detecting multiplier
PSM	Peak sampling multiplier
MPSM	Multiplexing peak sampling multiplier
SPSM	Switching peak sampling multiplier
PPRM	Pulse position responding multiplier
PPDM	Pulse position detecting multiplier
PPSM	Pulse position sampling multiplier
TDD	Time division divider
MTDD	Multiplexing time division divider
STDD	Switching time division divider
PRD	Peak responding divider
MPRD	Multiplexing peak responding divider
SPRD	Switching peak responding divider
PDD	Peak detecting divider
MPDD	Multiplexing peak detecting divider
SPDD	Switching peak detecting divider
PSD	Peak sampling divider
MPSD	Multiplexing peak sampling divider
SPSD	Switching peak sampling divider
PPRD	Pulse position responding divider
PPDD	Pulse position detecting divider
PPSD	Pulse position sampling divider
TDMCD	Time division multiplier cum divider
MTDMCD	Multiplexing time division multiplier cum divider

STDMCD	Switching time division multiplier cum divider
PRMCD	Peak responding multiplier cum divider
MPRMCD	Multiplexing peak responding multiplier cum divider
SPRMCD	Switching peak responding multiplier cum divider
PDMCD	Peak detecting multiplier cum divider
MPDMCD	Multiplexing peak detecting multiplier cum divider
SPDMCD	Switching peak detecting multiplier cum divider
PSMCD	Peak sampling multiplier cum divider
MPSMCD	Multiplexing peak sampling multiplier cum divider
SPSMCD	Switching peak sampling multiplier cum divider
PPRMCD	Pulse position responding multiplier cum divider
PPDMCD	Pulse position detecting multiplier cum divider
PPSMCD	Pulse position sampling multiplier cum divider

Introduction to the 555 Timer

Figure 0.1 shows the functional diagram of the 555 timer. The resistors R_1, R_2, and R_3 are used as voltage dividers and provide voltage references (1) $2V_{CC}/3$ for the upper comparator CMP_1 and (2) $V_{CC}/3$ for the lower comparator CMP_2.

Initially when the power supply is switched on, the output of the upper comparator CMP_1 will be LOW, i.e., R = 0, and the output of the lower comparator CMP_2 will be HIGH, i.e., S = 1. The flip flop outputs are Q = 1 and Q′ = 0. The timer output at pin 3 will be HIGH, transistor Q_1 is OFF, and hence the discharge pin 7 is at the open position.

With the threshold pin 6 and trigger pin 2 tied together, a rising voltage is applied to these connected pins 2 and 6. When the rising voltage is increased above 2 $V_{CC}/3$, the output of the upper comparator CMP_1 becomes HIGH, i.e., R = 1, and the output of the lower comparator CMP_2 becomes LOW, i.e., S = 0. The flip flop outputs are Q = 0 and Q′ = 1. The timer output at pin 3 will be LOW, transistor Q1 is ON, and hence the discharge pin 7 is at GND potential.

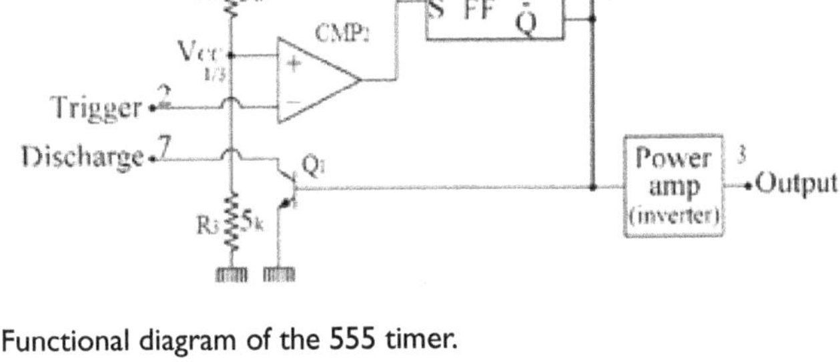

Figure 0.1 Functional diagram of the 555 timer.

Table 0.1 States of the 555 Timer

Sl. No.	*Trigger (pin 2)*	*Threshold (pin 6)*	*Output (pin 3)*	*Discharge (pin 7)*
1	Below $V_{CC}/3$	Below 2 $V_{CC}/3$	HIGH	OPEN
2	Below $V_{CC}/3$	Above 2 $V_{CC}/3$	Last state remains	Last state remains
3	Above $V_{CC}/3$	Below 2 $V_{CC}/3$	Last state remains	Last state remains
4	Above $V_{CC}/3$	Above 2 $V_{CC}/3$	LOW	GROUND

Now let us change the rising voltage in to a falling voltage. When the falling voltage goes below 1/3 VCC, the output of the upper comparator CMP_1 becomes LOW, i.e., R = 0, the output of lower comparator CMP_2 becomes HIGH, i.e., S = 1. The flip flop outputs are Q = 1 and Q′ = 0. The timer output at pin 3 will be HIGH, transistor Q_1 is OFF, and hence the discharge pin 7 is at the open position.

The reset pin 4 is used to reset the flip flop if there are any overrides in the operation. The transistor Q_2 is working as a buffer to isolate the reset input from the flip flop and transistor Q_1. The transistor Q_2 is driven by an internal reference voltage V_{REF} obtained from V_{CC}. The different operation states of the 555 timer are shown in Table 0.1.

Chapter 1

Time Division Multipliers—Multiplexing

1.1 SAW TOOTH WAVE BASED TIME DIVISION MULTIPLIERS

The circuit diagrams of saw tooth wave based time division multipliers are shown in Figure 1.1, and their associated waveforms are shown in Figure 1.2. A saw tooth wave V_{S1} of peak value V_R and time period T is generated by the 555 timer.

In the circuits of Figure 1.1, the comparator OA_2 compares the saw tooth wave V_{S1} of peak value V_R with the input voltage V_1 and produces a rectangular waveform V_M at its output. The ON time δ_T of this rectangular waveform V_M is given as

$$\delta_T = \frac{V_1}{V_R} T \tag{1.1}$$

The rectangular pulse V_M controls the multiplexer M_1. When V_M is HIGH, another input voltage V_2 is connected to the R_3C_2 low pass filter ('ay' is connected to 'a'). When V_M is LOW, zero voltage is connected to the R_3C_2 low pass filter ('ax' is connected to 'a'). Another rectangular pulse V_N with maximum value of V_2 is generated at the multiplexer M_1 output. The R_3C_2 low pass filter gives the average value of this pulse train V_N and is given as

$$V_O = \frac{1}{T} \int_0^{\delta_T} V_2 dt \tag{1.2}$$

$$V_O = \frac{V_2}{T} \delta_T \tag{1.3}$$

Equation (1.1) in (1.3) gives

$$V_O = \frac{V_1 V_2}{V_R} \tag{1.4}$$

where $V_R = 2/3\ V_{CC}$.

DOI: 10.1201/9781003362968-1

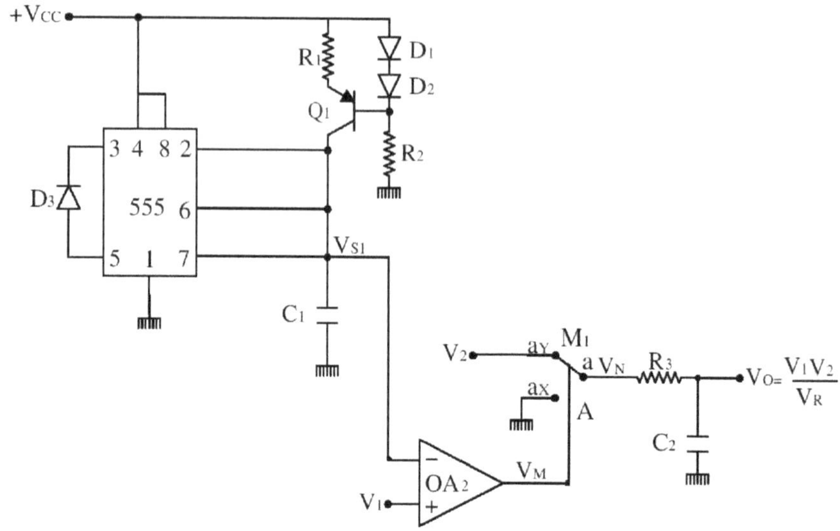

(a)

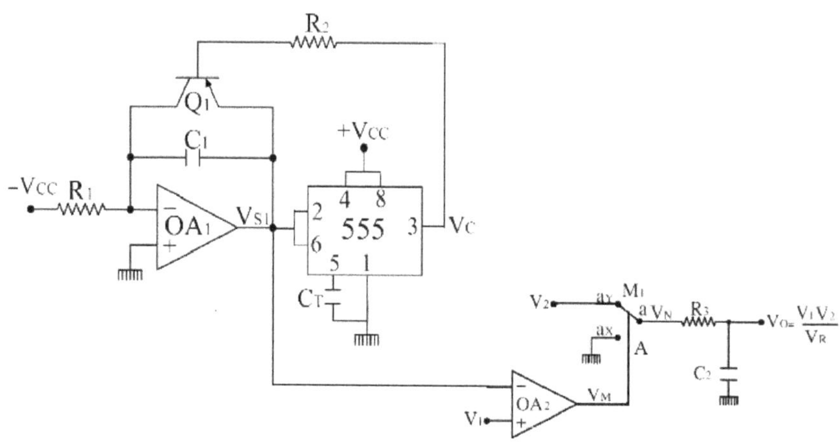

(b)

Figure 1.1 (a) Saw tooth wave based time division multiplier—type I. (b) Saw tooth wave based time division multiplier—type II.

1.2 TRIANGULAR WAVE REFERENCED TIME DIVISION MULTIPLIERS

The circuit diagrams of triangular wave based multipliers are shown in Figure 1.3, and their associated waveforms are shown in Figure 1.4. In Figure 1.3(a), a triangular wave V_{T1} with $\pm V_T$ peak to peak value and time period T is generated by the 555 timer.

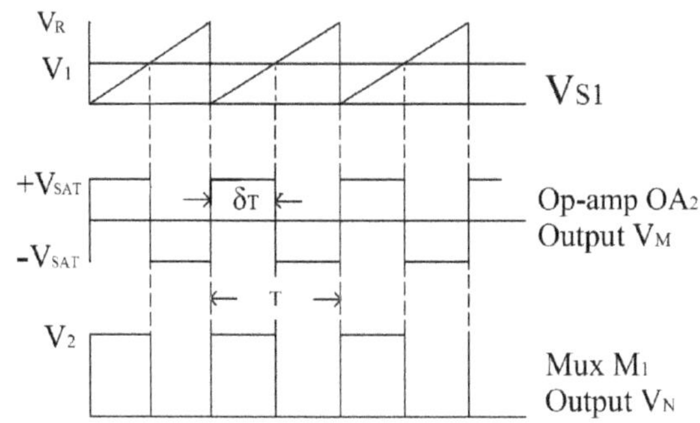

Figure 1.2 Associated waveforms of Figure 1.1.

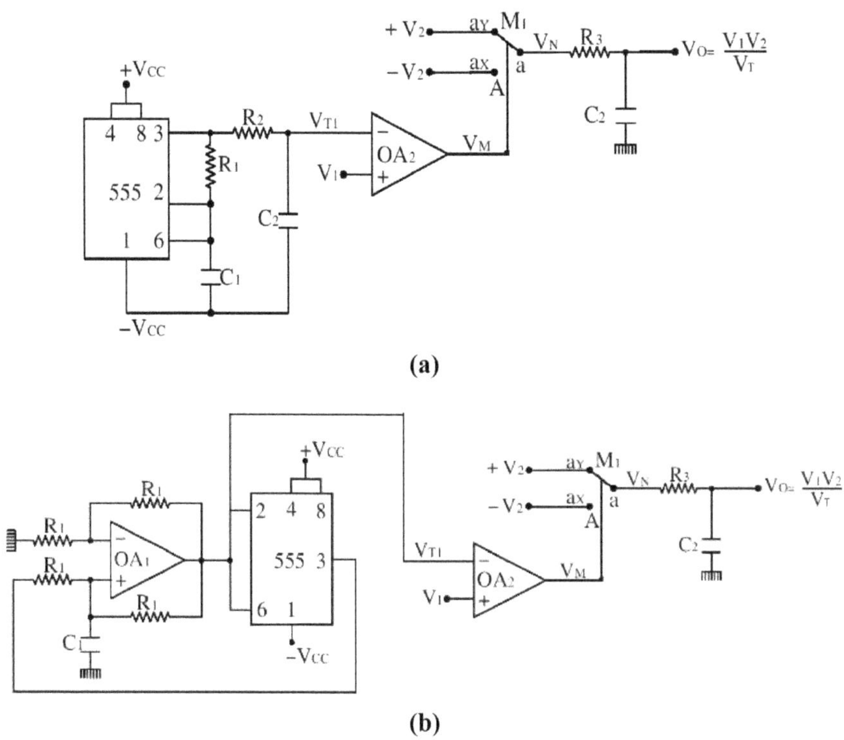

Figure 1.3 (a) Triangular wave based multiplier—type I. (b) Triangular wave based multiplier—type II.

One input voltage V_1 is compared with the generated triangular wave V_{T1} by the comparator on OA_2. An asymmetrical rectangular waveform V_M is generated at the comparator OA_2 output. From the waveforms shown in Figure 1.4, it is observed that

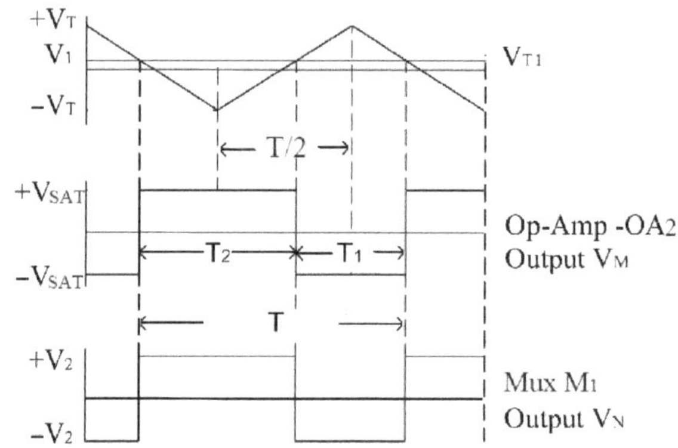

Figure 1.4 Associated waveforms of Figure 1.3(a) and (b).

$$T_1 = \frac{V_T - V_1}{2V_T}T,\ T_2 = \frac{V_T + V_1}{2V_T}T,\ T = T_1 + T_2 \tag{1.5}$$

This rectangular wave V_M is given as control input to the multiplexer M_1. The multiplexer M_1 connects the other input voltage $+V_2$ during T_2 ('ay' is connected to 'a') and $-V_2$ during T_1 ('ax' is connected to 'a'). Another rectangular asymmetrical wave V_N with peak to peak value of $\pm V_2$ is generated at the multiplexer M_1 output. The R_3C_2 low pass filter gives the average value of the pulse train V_N, which is given as

$$V_O = \frac{1}{T}\left[\int_O^{T_2} V_2\,dt + \int_{T_2}^{T_1+T_2} (-V_2)\,dt\right] = \frac{V_2}{T}(T_2 - T_1) \tag{1.6}$$

Equation (1.5) in (1.6) gives

$$V_O = \frac{V_1 V_2}{V_T} \tag{1.7}$$

where $V_T = V_{CC}/3$. (1.8)

1.3 TIME DIVISION MULTIPLIER WITH NO REFERENCE—TYPE I

The multipliers using the time division principle without using any reference clock is shown in Figure 1.5, and its associated waveforms are shown in Figure 1.6.

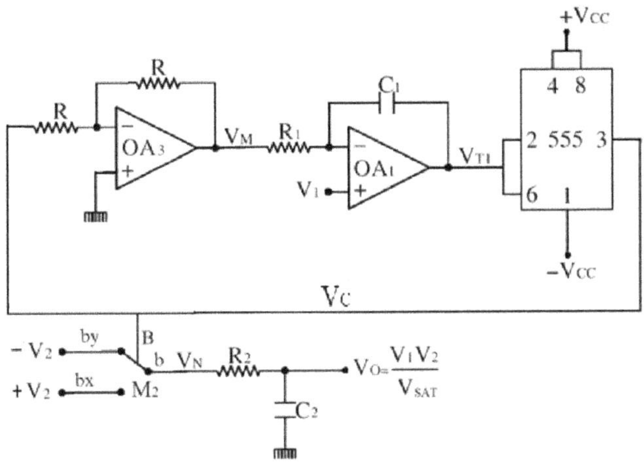

Figure 1.5 Time division multiplier without reference clock—type I.

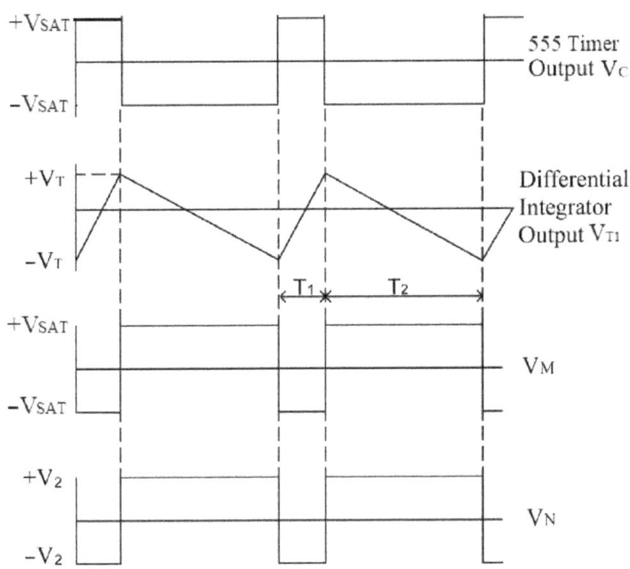

Figure 1.6 Associated waveforms of Figure 1.5.

Initially when the 555 timer output is HIGH, the inverting amplifier OA_3 gives $-V_{SAT}$ to the differential integrator composed by resistor R_1, capacitor C_1, and op-amp OA_1. The output of the differential integrator will be

$$V_{T1} = \frac{1}{R_1C_1}\int(V_1 + V_{SAT})dt$$

$$V_{T1} = \frac{(V_{SAT} + V_1)}{R_1C_1}t \tag{1.9}$$

The output of the differential integrator rises toward positive saturation, and when it reaches the voltage level of $+V_T$, the 555 timer output becomes LOW. The inverting amplifier OA_3 gives $+V_{SAT}$ to the differential integrator composed of resistor R_1, capacitor C_1 and op-amp OA_1. Now the output of the differential integrator will be

$$V_{T1} = \frac{1}{R_1C_1}\int(V_1 - V_{SAT})dt$$

$$V_{T1} = -\frac{(V_{SAT} - V_1)}{R_1C_1}t \tag{1.10}$$

The output of the differential integrator reverses toward negative saturation, and when it reaches the voltage level $-V_T$, the 555 timer output becomes HIGH, and the cycle therefore repeats, to give an asymmetrical rectangular wave V_C at the output of the 555 timer.

$$V_T = \frac{V_{CC}}{3} \tag{1.11}$$

From the waveforms shown in Figure 1.6, it is observed that

$$T_1 = \frac{V_{SAT} - V_1}{2V_{SAT}}T,\ T_2 = \frac{V_{SAT} + V_1}{2V_{SAT}}T,\ T = T_1 + T_2 \tag{1.12}$$

The asymmetrical rectangular wave V_C controls the multiplexer M_2. The multiplexer M_2 connects $+V_2$ during the OFF time V_C ('bx' is connected to 'b') and $-V_2$ during the ON time of the rectangular wave V_C ('by' is connected to 'b'). Another rectangular wave V_N is generated at the multiplexer M_2 output. The R_2C_2 low pass filter gives the average value of this pulse train V_N and is given as

$$V_O = \frac{1}{T}\left[\int_O^{T_2} V_2\,dt + \int_{T_2}^{T_1+T_2}(-V_2)dt\right]$$

$$V_O = \frac{V_2(T_2 - T_1)}{T} \tag{1.13}$$

Equation (1.12) in (1.13) gives

$$V_O = \frac{V_1 V_2}{V_{SAT}} \tag{1.14}$$

1.4 TIME DIVISION MULTIPLIER NO REFERENCE—TYPE II

The time division multiplier using the time division principle without using any reference clock is shown in Figure 1.7, and its associated waveforms are shown in Figure 1.8.

Initially the 555 timer output V_C is HIGH. $-V_1$ is connected to the differential integrator by the multiplexer M_1 ('ay' is connected to 'a'). The inverting amplifier OA_3 output will be LOW, i.e., $-V_{SAT}$. The output of the differential integrator will be

$$V_{T1} = \frac{1}{R_1 C_1} \int (V_O + V_1)dt$$

$$V_{T1} = \frac{(V_O + V_1)}{R_1 C_1} t \tag{1.15}$$

The output of the differential integrator rises toward positive saturation, and when it reaches the voltage level of $+V_T$, the 555 timer output becomes LOW. $+V_1$ is connected to the differential integrator by the multiplexer M_1

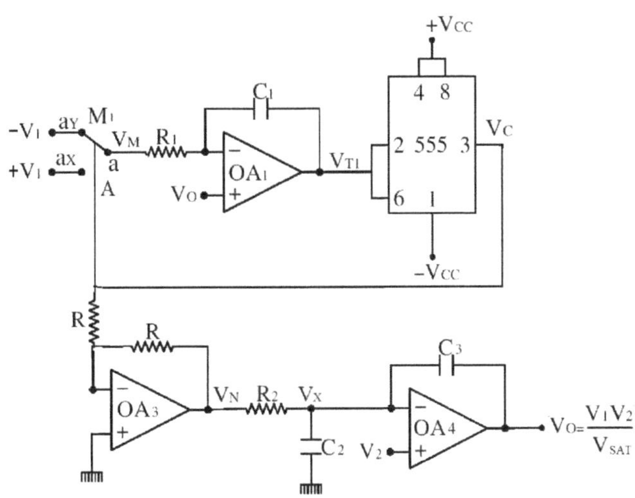

Figure 1.7 Time division multiplier without reference clock—type II.

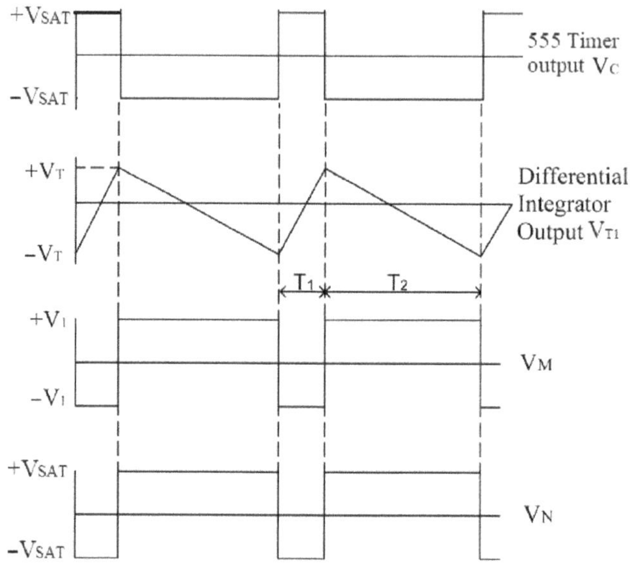

Figure 1.8 Associated waveforms of Figure 1.7.

('ax' is connected to 'a'). The inverting amplifier OA_3 output will be HIGH, i.e., $+V_{SAT.}$ Now the output of the differential integrator will be

$$V_{T1} = \frac{1}{R_1C_1}\int (V_O - V_1)dt$$

$$V_{T1} = -\frac{(V_1 - V_O)}{R_1C_1}t \qquad (1.16)$$

The output of the differential integrator reverses toward negative saturation, and when it reaches the voltage level $-V_T$, the 555 timer output becomes HIGH, and the cycle therefore repeats, to give an asymmetrical rectangular wave V_C at the output of the 555 timer.

$$V_T = \frac{V_{CC}}{3} \qquad (1.17)$$

From the waveforms shown in Figure 1.8, it is observed that

$$T_1 = \frac{V_1 - V_O}{2V_1}T,\ T_2 = \frac{V_1 + V_O}{2V_1}T,\ T = T_1 + T_2 \qquad (1.18)$$

Another rectangular wave V_N with $\pm V_{SAT}$ results as the peak to peak value is generated at the inverting amplifier OA_3 output. The R_2C_2 low pass filter gives the average value of this pulse train V_N and is given as

$$V_X = \frac{1}{T}\left[\int_{O}^{T_2} V_{SAT}\, dt + \int_{T_2}^{T_1+T_2} (-V_{SAT})\, dt\right]$$

$$V_X = \frac{V_{SAT}(T_2 - T_1)}{T} \tag{1.19}$$

Equations (1.18) in (1.19) gives

$$V_X = \frac{V_O V_{SAT}}{V_1} \tag{1.20}$$

The op-amp OA_4 is at the negative closed loop configuration, and a positive dc voltage is ensured in the feedback loop. Hence its non-inverting terminal voltage is equal to its inverting terminal voltage, i.e.,

$$V_2 = V_X \tag{1.21}$$

From equations (1.20) and (1.21)

$$V_O = \frac{V_1 V_2}{V_{SAT}} \tag{1.22}$$

1.5 TIME DIVISION MULTIPLIER USING 555 ASTABLE MULTIVIBRATOR

1.5.1 Time Division Multiplier Using 555 Astable Multivibrator—Type I

The circuit diagram of the multiplier using the 555 timer astable multivibrator is shown in Figure 1.9, and its associated waveforms are shown in Figure 1.10. Refer to the internal diagram of the 555 timer IC shown in Figure 0.1. Initially when we switch on the power supply, the output of the upper comparator CMP_1 will be LOW, i.e., R = 0, the output of the lower comparator CMP_2 will be HIGH, i.e., S = 1. The flip flop outputs are Q = 1 and Q′ = 0. The timer output at pin 3 will be HIGH, transistor Q_1 is OFF, and hence the discharge pin 7 is at the open position.

The capacitor C_1 is charging toward $+V_{CC}$ through the resistors R_1 and R_2 with a time constant of $(R_1+R_2)C_1$, and its voltage is rising exponentially.

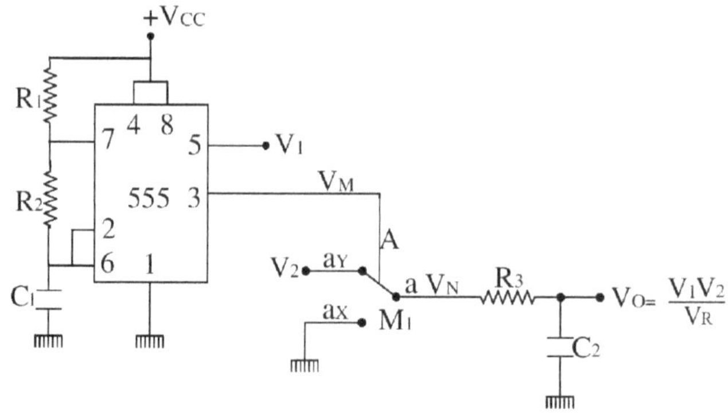

Figure 1.9 Multiplier with 555 timer astable multivibrator.

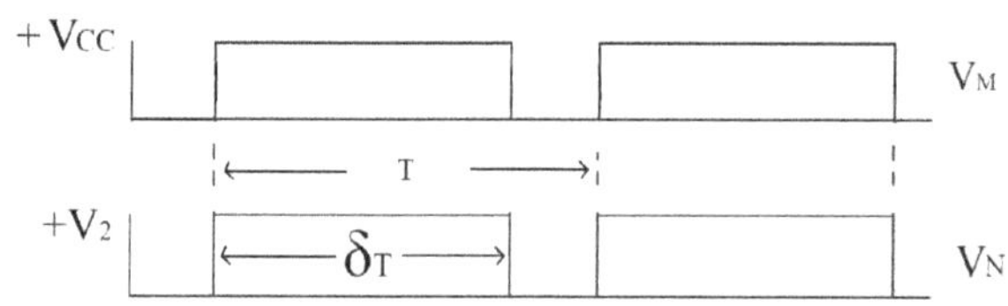

Figure 1.10 Associated waveforms of Figure 1.9.

When the capacitor voltage is rising above the voltage V_1, the output of the upper comparator CMP_1 becomes HIGH, i.e., R = 1, and the output of lower comparator CMP_2 becomes LOW, i.e., S = 0. The flip flop outputs are Q = 0 and Q′ = 1. The timer output at pin 3 will be LOW, transistor Q_1 is ON and hence the discharge pin 7 is at GND potential. Now the capacitor C_1 is discharging to GND potential through the resistor R_2 with a time constant of R_2C. When the capacitor voltage falls below 1/3 V_{CC}, the output of the upper comparator CMP_1 becomes LOW, i.e., R = 0, the output of the lower comparator CMP_2 becomes HIGH, i.e., S = 1. The flip flop outputs are Q = 1 and Q′ = 0. The timer output at pin 3 will be HIGH, transistor Q_1 is OFF, and hence the discharge pin 7 is at the open position.

Now the capacitor starts charging toward $+V_{CC}$, and the cycle therefore repeats to produce periodic pulses at the output pin 3 of the 555 timer.

The ON time of the 555 timer output V_M is proportional to V_1, which is applied at its pin 5. The pulse V_M controls the multiplexer M_1. During the ON time δ_T, the second input voltage V_2 is connected to the R_3C_2 low pass filter ('ay' is connected to 'a'). During the OFF time of V_M, zero voltage is connected to the R_3C_2 low pass filter ('ax' is connected to 'a'). Another rectangular waveform V_N, with V_2 as the peak value, is generated at the output of the multiplexer M_1.

$$\delta_T = \frac{V_1}{V_R} T \tag{1.23}$$

The R_3C_2 low pass filter gives the average value of this pulse train V_N and is given as

$$V_O = \frac{1}{T}\int_0^{\delta_T} V_2 dt = \frac{V_2}{T}\delta_T$$

$$V_O = \frac{V_1 V_2}{V_R} \tag{1.24}$$

where V_R is a constant value.

1.5.2 Multiplier from 555 Astable Multivibrator—Type II

The circuit diagram of the divider using the 555 astable multivibrator is shown in Figure 1.11, and its associated waveforms are shown in Figure 1.12. Refer to the internal diagram of the 555 timer IC shown in Figure 0.1. Initially when we switch on the power supply, the output of the upper comparator CMP_1 will be LOW, i.e., R = 0, the output of the lower comparator CMP_2 will be HIGH, i.e., S = 1. The flip flop outputs are Q = 1 and Q′ = 0. The timer output at pin 3 will be HIGH, transistor Q_1 is OFF, and hence the discharge pin 7 is at the open position.

The capacitor C_1 is charging toward V_1 through the resistors R_1 and R_2 with a time constant of $(R_1+R_2)C_1$, and its voltage is rising exponentially.

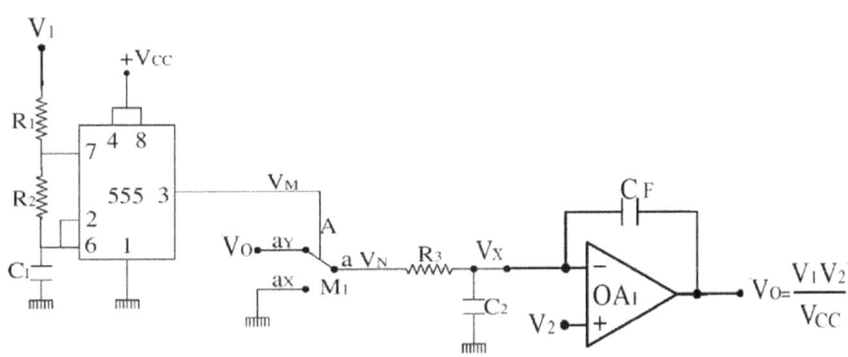

Figure 1.11 Multiplier from 555 astable—type II.

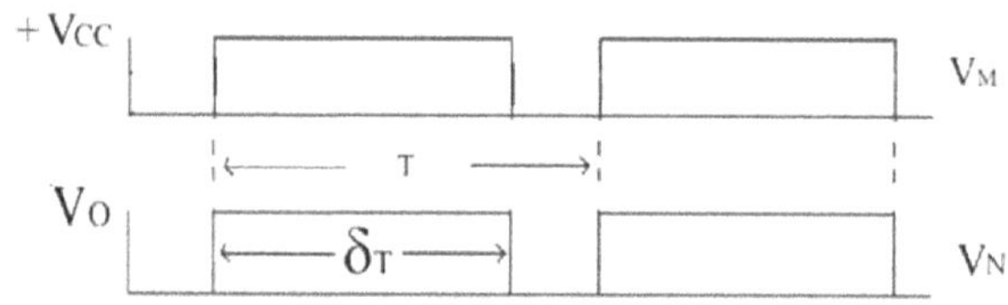

Figure 1.12 Associated waveforms of Figure 1.11.

When the capacitor voltage is rising above the voltage 2/3 V_{CC}, the output of the upper comparator CMP_1 becomes HIGH, i.e., R = 1, and the output of the lower comparator CMP_2 becomes LOW, i.e., S = 0. The flip flop outputs are Q = 0 and Q′ = 1. The timer output at pin 3 will be LOW, transistor Q_1 is ON and hence the discharge pin 7 is at GND potential. Now the capacitor C_1 is discharging to GND potential through the resistor R_2 with a time constant of R_2C. When the capacitor voltage falls below 1/3 V_{CC}, the output of the upper comparator CMP_1 becomes LOW, i.e., R = 0, and the output of the lower comparator CMP_2 becomes HIGH, i.e., S = 1. The flip flop outputs are Q = 1 and Q′ = 0. The timer output at pin 3 will be HIGH, transistor Q_1 is OFF, and hence the discharge pin 7 is at the open position.

Now the capacitor starts charging toward V_1, and the cycle therefore repeats to produce periodic pulses at the output pin 3 of the 555 timer.

The ON time of the 555 timer output V_M is inversely proportional to V_O. The 555 timer output controls the multiplexer M_1. During the ON time δ_T, the input voltage Vo is connected to the R_3C_2 low pass filter ('ay' is connected to 'a'). During the OFF time of V_M, zero voltage is connected to the R_3C_2 low pass filter ('ax' is connected to 'a'). Another rectangular waveform V_N, with V_O as the peak value, is generated at the output of multiplexer M_1. The ON time δ_T of this rectangular pulse V_N is given as

$$\delta_T = \frac{V_R}{V_1} T \tag{1.25}$$

where V_R is a constant value.

The R_3C_2 low pass filter gives the average value of this pulse train V_N and is given as

$$V_X = \frac{1}{T}\int_0^{\delta_T} V_O dt = \frac{V_O}{T}\delta_T$$

$$V_X = \frac{V_O}{V_1} V_R \tag{1.26}$$

The op-amp OA_1 is kept in a negative closed loop configuration, and a positive dc voltage is ensured in the feedback. Hence its inverting terminal voltage will be equal to its non-inverting terminal voltage, i.e.,

$$V_X = V_2 \tag{1.27}$$

From equations (1.26) and (1.27),

$$V_O = \frac{V_1 V_2}{V_R} \tag{1.28}$$

1.6 MULTIPLIER FROM 555 MONOSTABLE MULTIVIBRATOR

1.6.1 Multiplier from 555 Monostable Multivibrator—Type I

The circuit diagram of a multiplier using the 555 timer monostable multivibrator is shown in Figure 1.13, and its associated waveforms are shown in Figure 1.14. Refer to the internal diagram of the 555 timer IC shown in

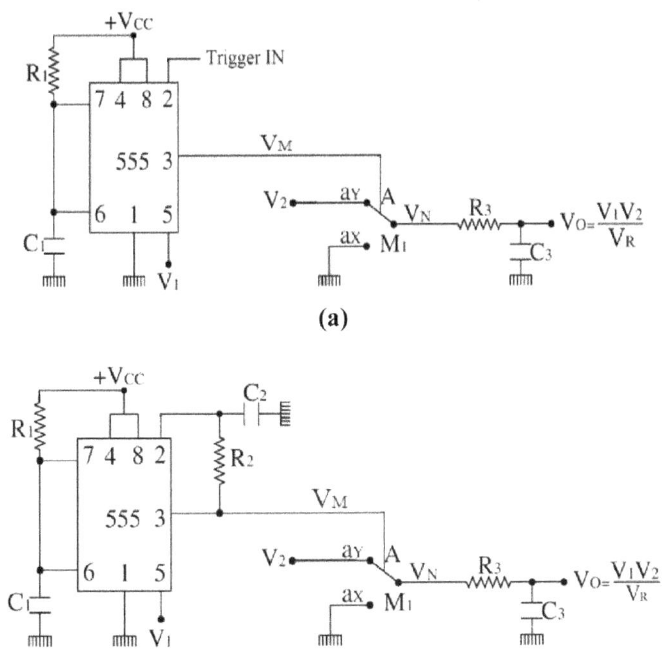

Figure 1.13 (a) Multiplier from 555 monostable. (b) Multiplier with 555 re-trigger monostable multivibrator.

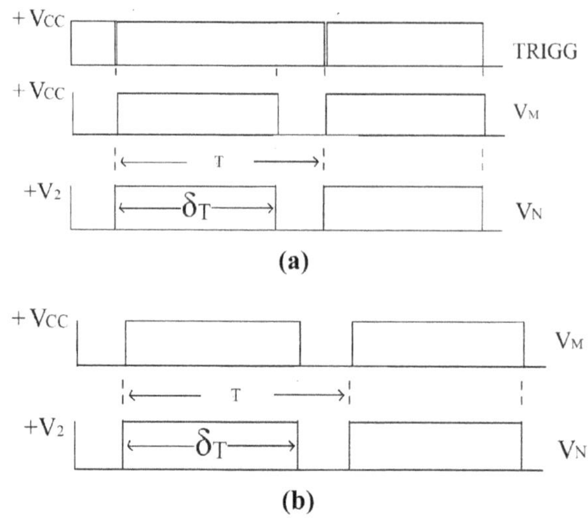

Figure 1.14 (a) Associated waveforms of Figure 1.13(a). (b) Associated waveforms of Figure 1.13(b).

Figure 0.1. Initially when the power supply is switched on, the output of the upper comparator CMP_1 will be LOW, i.e., R = 0, and the output of the lower comparator CMP_2 will be HIGH, i.e., S = 1. The flip flop outputs are Q = 1 and Q′ = 0. The timer output at pin 3 will be HIGH, transistor Q_1 is OFF, and hence the discharge pin 7 is at the open position. The capacitor C_1 is charging toward $+V_{CC}$ through the resistor R_1. The capacitor voltage is rising exponentially and when it reaches the value of V_1, the output of the upper comparator CMP_1 becomes HIGH, i.e., R = 1, and the output of the lower comparator CMP_2 becomes LOW, i.e., S = 0. The flip flop outputs are Q = 0 and Q′ = 1. The timer output at pin 3 will be LOW, transistor Q_1 is ON, and hence the discharge pin 7 is at GND potential. Now the capacitor C_1 is short circuited, zero volts is existing at pin 6, the output of the upper comparator CMP_1 becomes LOW, i.e., R = 0. A trigger pulse is applied at pin 2, and when the trigger voltage comes down to 1/3 V_{CC}, the output of the lower comparator CMP_2 becomes HIGH, i.e., S = 1. The flip flop outputs are Q = 1 and Q′ = 0. The timer output at pin 3 will be HIGH, transistor Q_1 is OFF, and hence the discharge pin 7 is at the open position.

Now the capacitor C_1 is charging toward $+V_{CC}$, and the sequence therefore repeats for every trigger input pulse.

The ON time of the 555 timer output V_M is proportional to V_1, which is applied at its pin 5. The 555 timer output controls the multiplexer M_1. During the ON time δ_T, the second input voltage V_2 is connected to the R_3C_3 low pass filter ('ay' is connected to 'a'). During the OFF time of V_M, zero voltage is connected to the R_3C_3 low pass filter ('ax' is connected to 'a'). Another rectangular waveform V_N, with V_2 as the peak value, is generated at the output of multiplexer M_1. The ON time δ_T of this rectangular waveform V_N is given as

$$\delta_T = \frac{V_1}{V_R} T \tag{1.29}$$

The R_3C_3 low pass filter gives the average value of this pulse train V_N and is given as

$$V_O = \frac{1}{T}\int_0^{\delta_T} V_2 dt = \frac{V_2}{T}\delta_T$$

$$V_O = \frac{V_1 V_2}{V_R} \tag{1.30}$$

where V_R is a constant value.

The multiplier using the 555 re-trigger monostable multivibrator is shown in Figure 1.13(b).

1.6.2 Multiplier from 555 Monostable Multivibrator—Type II

The circuit diagram of a divider using the 555 monostable multivibrator is shown in Figure 1.15, and its associated waveforms are shown in Figure 1.16. Refer to the internal diagram of the 555 timer IC shown in Figure 0.1. Initially when the power supply is switched on, the output of the upper comparator CMP_1 will be LOW, i.e., R = 0, and the output of the lower comparator CMP_2 will be HIGH, i.e., S = 1. The flip flop outputs are Q = 1 and Q′ = 0. The timer output at pin 3 will be HIGH, transistor Q_1 is OFF, and hence the discharge pin 7 is at the open position. The capacitor C_1 is charging toward V_1 through the resistor R_1. The capacitor voltage is rising exponentially, and when it reaches the value of 2/3 V_{CC}, the output of the upper comparator CMP_1 becomes HIGH, i.e., R = 1, and the output of the lower comparator CMP_2 becomes LOW, i.e., S = 0. The flip flop outputs are Q = 0 and Q′ = 1. The timer output at pin 3 will be LOW, transistor Q_1 is ON, and hence the discharge pin 7 is at GND potential. Now the capacitor C_1 is short circuited, zero voltage exists at pin 6, and the output of the upper comparator CMP_1 becomes LOW, i.e., R = 0. A trigger pulse is applied at pin 2, and when the trigger voltage comes down to 1/3 V_{CC}, the output of the lower comparator CMP_2 becomes HIGH, i.e., S = 1. The flip flop outputs are Q = 1 and Q′ = 0. The timer output at pin 3 will be HIGH, transistor Q_1 is OFF, and hence the discharge pin 7 is at the open position.

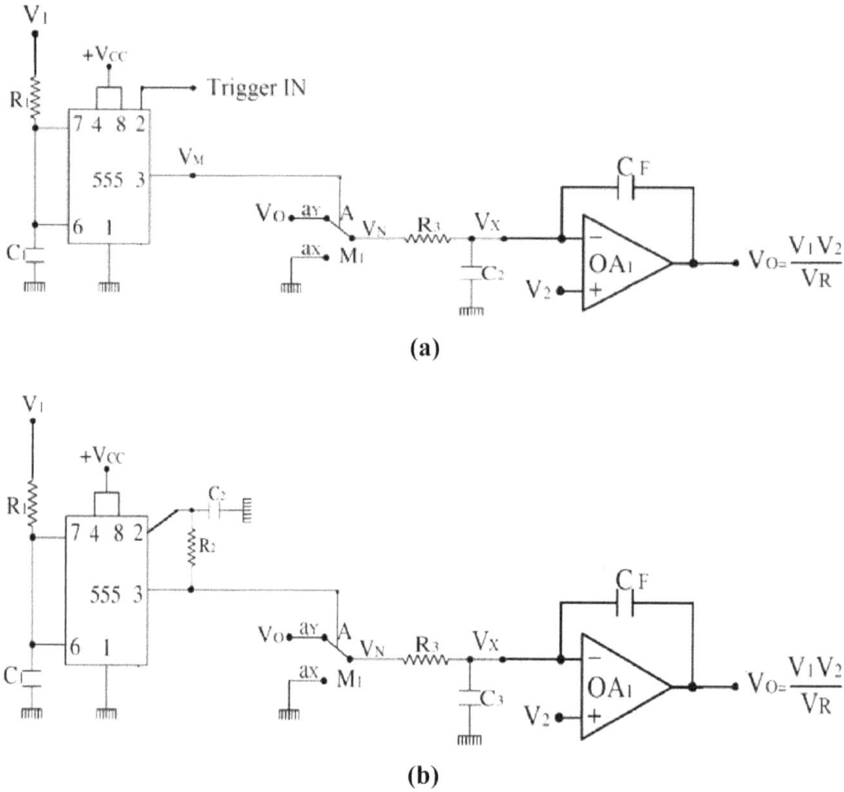

Figure 1.15 (a) Multiplier using 555 timer monostable multivibrator. (b) Multiplier using re-trigger monostable multivibrator.

Now the capacitor C_1 is charging toward V_1, and the sequence therefore repeats for every trigger input pulse.

The ON time of the 555 timer output V_M is inversely proportional to V_1. The output of the 555 timer controls the multiplexer M_1. During the ON time δ_T, the voltage V_O is connected to the R_3C_3 low pass filter ('ay' is connected to 'a'). During the OFF time of V_M, zero voltage is connected to the R_3C_3 low pass filter ('ax' is connected to 'a'). Another rectangular waveform V_N, with V_O as peak value, is generated at the output of the multiplexer M_1.

$$\delta_T = \frac{V_R}{V_1} T \tag{1.31}$$

The R_3C_3 low pass filter gives the average value of this pulse train V_N and is given as

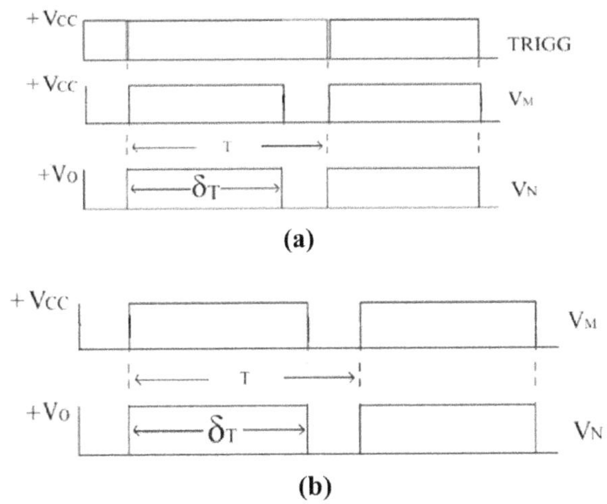

Figure 1.16 (a) Associated waveforms of Figure 1.15(a). (b) Associated waveforms of Figure 1.15(b).

$$V_X = \frac{1}{T}\int_0^{\delta_T} V_O dt = \frac{V_O}{T}\delta_T$$

$$V_X = \frac{V_O}{V_1} V_R \tag{1.32}$$

where V_R is a constant value.

The op-amp OA_1 is kept in a negative closed loop configuration, and a positive dc voltage is ensured in the feedback. Hence its inverting terminal voltage will be equal to its non-inverting terminal voltage, i.e.,

$$V_X = V_2 \tag{1.33}$$

From equations (1.32) and (1.33),

$$V_O = \frac{V_1 V_2}{V_R} \tag{1.34}$$

Figure 1.15(b) shows the re-trigger monostable multivibrator used as an analog multiplier.

Chapter 2

Time Division Multipliers—Switching

If the width of a pulse train is made proportional to one voltage and the amplitude of the same pulse train to a second voltage, then the average value of this pulse train is proportional to the product of two voltages and is called a time division multiplier, a pulse averaging multiplier, or a sigma delta multiplier. The time division multiplier can be implemented using (1) a triangular wave, (2) a saw tooth wave, and (3) no reference wave.

There are two types of time division multipliers (TDM) (1) multiplexing TDM (MTDM) and (2) switching TDM (STDM). A time division multiplier using analog 2 to 1 multiplexers is called a multiplexing TDM. A time division multiplier using analogue switches is called a switching TDM. Multiplexing time division multipliers are described in chapter 3, and switching time division multipliers are described in this chapter.

2.1 SAW TOOTH WAVE BASED TIME DIVISION MULTIPLIERS

The circuit diagrams of saw tooth wave based time division multipliers are shown in Figure 2.1, and their associated waveforms are shown in Figure 2.2. A saw tooth wave V_{S1} of peak value V_R and time period T is generated by the 555 timer.

In the circuits of Figure 2.1, the comparator OA_2 compares the saw tooth wave V_{S1} of peak value V_R with the input voltage V_1 and produces a rectangular waveform V_M at its output. The ON time δ_T of this rectangular waveform V_M is given as

$$\delta_T = \frac{V_1}{V_R} T \tag{2.1}$$

The rectangular pulse V_M controls the switch S_1. When V_M is HIGH, another input voltage V_2 is connected to the R_3C_2 low pass filter (switch S_1 is closed). When V_M is LOW, zero voltage is connected to the R_3C_2 low pass filter (switch S_1 is opened). Another rectangular pulse V_N with a maximum value

DOI: 10.1201/9781003362968-2

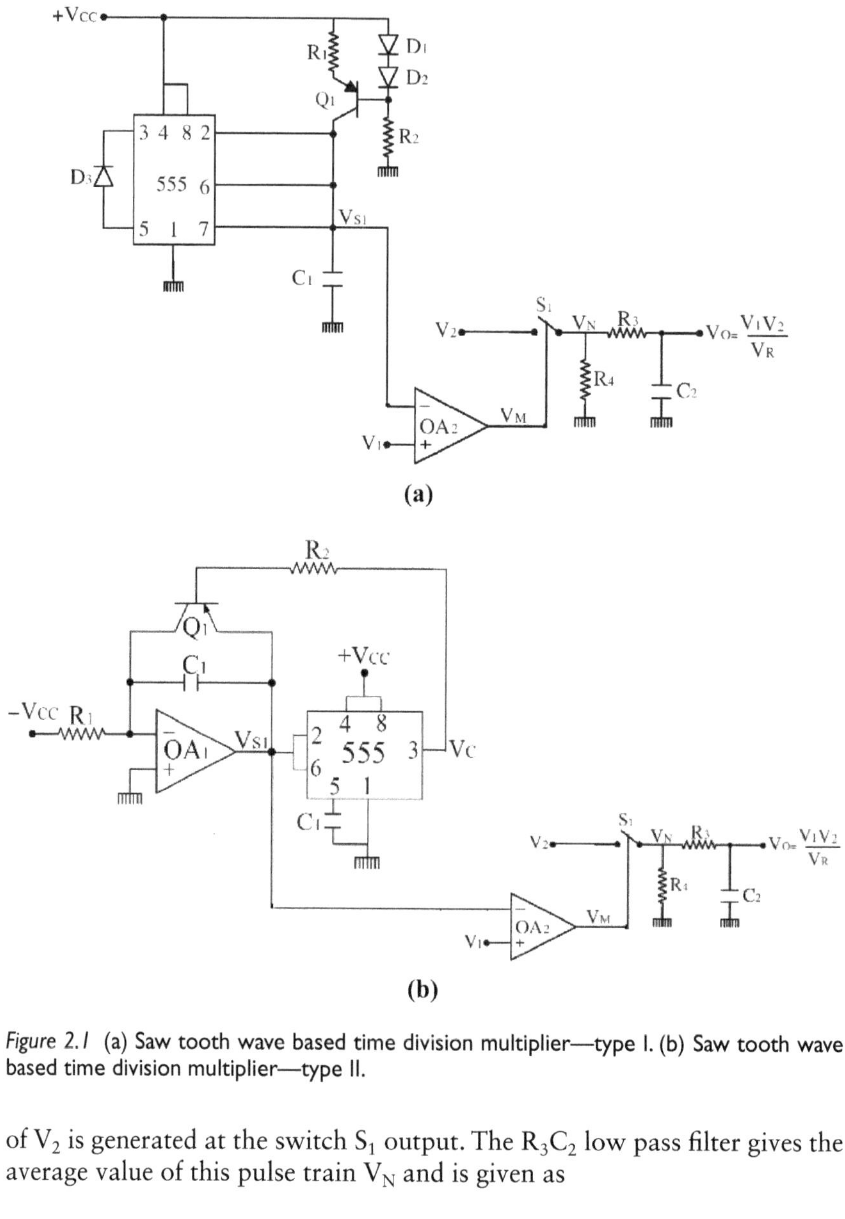

Figure 2.1 (a) Saw tooth wave based time division multiplier—type I. (b) Saw tooth wave based time division multiplier—type II.

of V_2 is generated at the switch S_1 output. The R_3C_2 low pass filter gives the average value of this pulse train V_N and is given as

$$V_O = \frac{1}{T}\int_0^{\delta_T} V_2 dt \tag{2.2}$$

$$V_O = \frac{V_2}{T}\delta_T \tag{2.3}$$

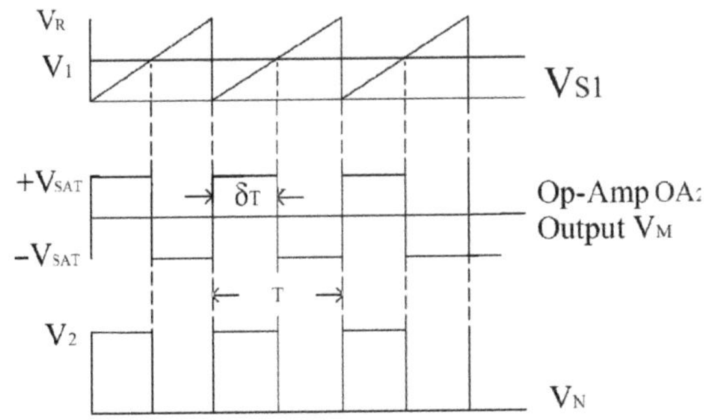

Figure 2.2 Associated waveforms of Figure 2.1.

Equation (2.1) in (2.3) gives

$$V_O = \frac{V_1 V_2}{V_R} \tag{2.4}$$

where $V_R = 2/3\ V_{CC}$.

2.2 TRIANGULAR WAVE REFERENCED TIME DIVISION MULTIPLIERS

The circuit diagrams of triangular wave based multipliers are shown in Figure 2.3, and their associated waveforms are shown in Figure 2.4. A triangular wave V_{T1} with $\pm V_T$ peak to peak value and time period T is generated by the 555 timer.

One input voltage V_1 is compared with the generated triangular wave V_{T1} by the comparator on OA_2. An asymmetrical rectangular waveform V_M is generated at the comparator OA_2 output. From the waveforms shown in Figure 2.5, it is observed that

$$T_1 = \frac{V_T - V_1}{2V_T}T,\ T_2 = \frac{V_T + V_1}{2V_T}T,\ T = T_1 + T_2 \tag{2.5}$$

This rectangular wave V_M is given as the control input to the switch S_1. During T_2 of V_M, the switch S_1 is closed, and the op-amp OA_3 will work as non-inverting amplifier. $+V_2$ will be its output, i.e., $V_N = +V_2$. During T_1 of V_M, the switch S_1 is opened, and the op-amp will work as inverting amplifier. $-V_2$ will be at its output, i.e., $V_N = -V_2$. Another rectangular asymmetrical wave V_N, with a peak to peak value of $\pm V_2$, is generated at the op-amp OA_3

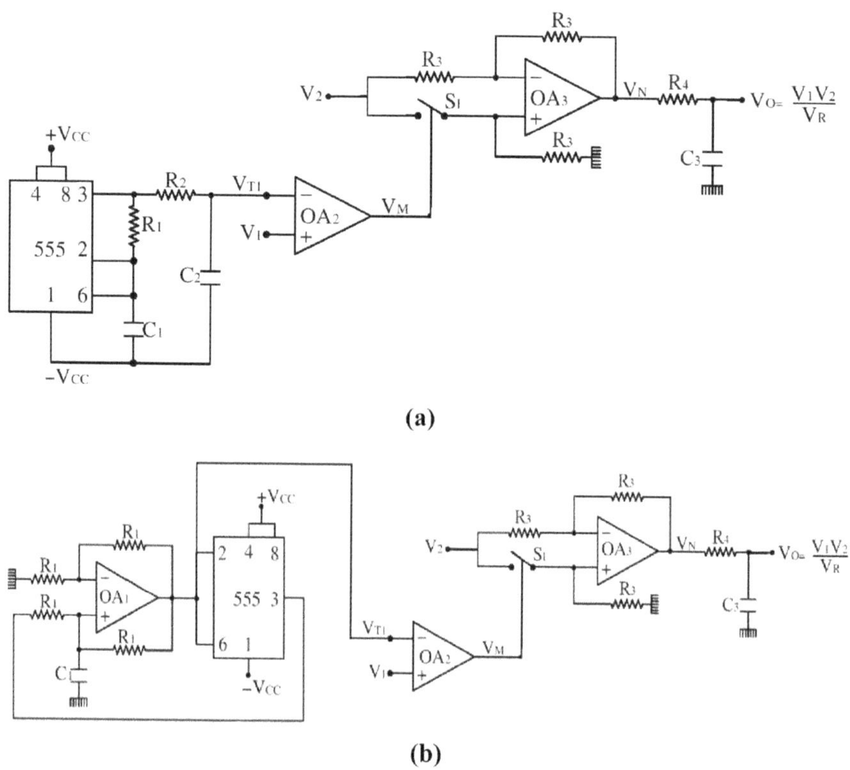

Figure 2.3 (a) Triangular wave based multiplier—type I. (b) Triangular wave based multiplier—type II.

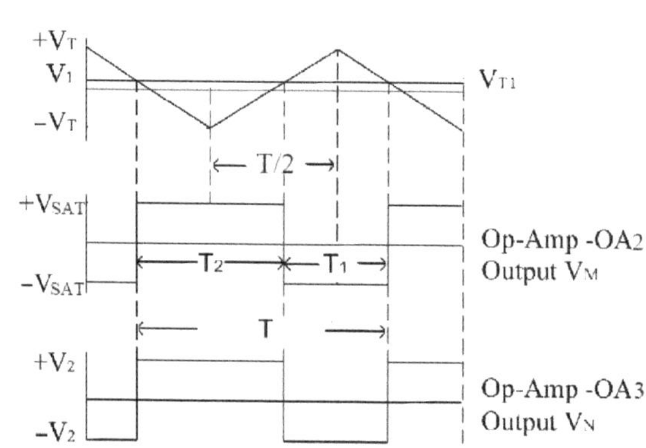

Figure 2.4 Associated waveforms of Figure 2.3(a) and (b).

output. The R_4C_3 low pass filter gives an average value of the pulse train V_N and is given as

$$V_O = \frac{1}{T}\left[\int_{O}^{T_2} V_2\,dt + \int_{T_2}^{T_1+T_2} (-V_2)\,dt\right] = \frac{V_2}{T}(T_2 - T_1) \tag{2.6}$$

Equation (2.5) in (2.6) gives

$$V_O = \frac{V_1 V_2}{V_T} \tag{2.7}$$

where $V_T = V_{CC}/3$. (2.8)

2.3 TIME DIVISION MULTIPLIER WITH NO REFERENCE—TYPE I

The multiplier using the time division principle without using any reference clock is shown in Figure 2.5, and its associated waveforms are shown in Figure 2.6.

Initially the 555 timer output is HIGH. The inverting amplifier OA_3 output is $-V_{SAT}$. The output of the differential integrator will be

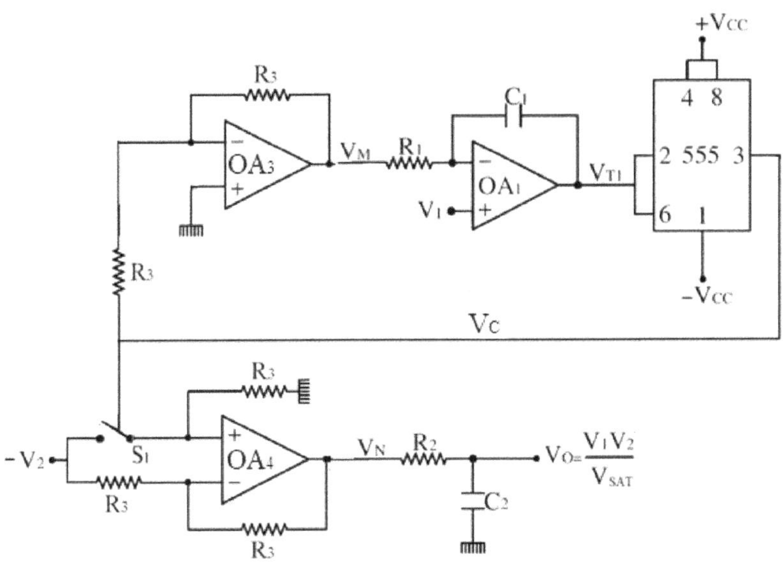

Figure 2.5 Time division multiplier without reference clock.

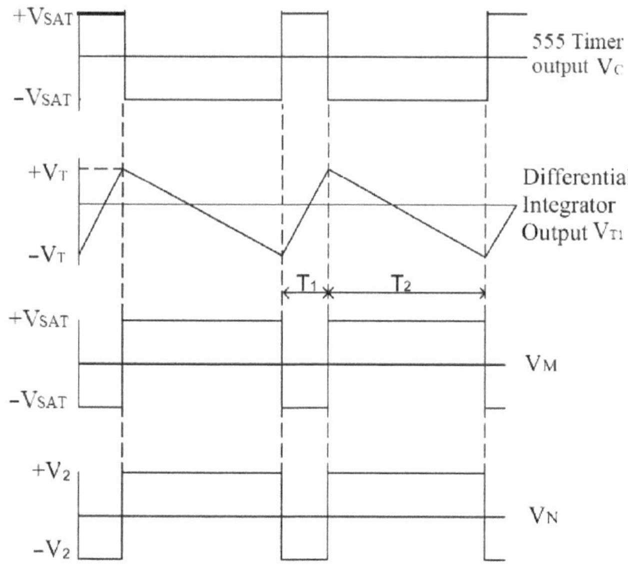

Figure 2.6 Associated waveforms of Figure 2.5.

$$V_{T1} = \frac{1}{R_1C_1}\int (V_1 + V_{SAT})dt$$

$$V_{T1} = \frac{(V_{SAT} + V_1)}{R_1C_1}t \qquad (2.9)$$

The output of the differential integrator rises toward positive saturation, and when it reaches the voltage level of $+V_T$, the 555 timer output becomes LOW. The inverting amplifier OA_3 output is $+V_{SAT}$. Now the output of the differential integrator will be

$$V_{T1} = \frac{1}{R_1C_1}\int (V_1 - V_{SAT})dt$$

$$V_{T1} = -\frac{(V_{SAT} - V_1)}{R_1C_1}t \qquad (2.10)$$

The output of the differential integrator reverses toward negative saturation, and when it reaches the voltage level $-V_T$, the 555 timer output becomes HIGH, and the cycle therefore repeats, to give an asymmetrical rectangular wave V_C at the output of the 555 timer.

$$V_T = \frac{V_{CC}}{3} \tag{2.11}$$

From the waveforms shown in Figure 2.6, it is observed that

$$T_1 = \frac{V_{SAT} - V_1}{2V_{SAT}} T,\ T_2 = \frac{V_{SAT} + V_1}{2V_{SAT}} T,\ T = T_1 + T_2 \tag{2.12}$$

The asymmetrical rectangular wave V_C controls switch S_1. The op-amp OA_4 gives $-V_2$ during the ON time T_1 of the rectangular waveform V_C (the switch S_1 is closed, and the op-amp OA_4 will work as a non-inverting amplifier) and $+V_2$ during the OFF time T_2 of the rectangular wave V_C (the switch S_1 is opened, and the op-amp OA_4 will work as an inverting amplifier). Another rectangular wave V_N with a peak to peak value of $\pm V_2$ is generated at the output of op-amp OA_4. The R_2C_2 low pass filter gives the average value of this pulse train V_N and is given as

$$V_O = \frac{1}{T}\left[\int_O^{T_2} V_2\,dt + \int_{T_2}^{T_1+T_2} (-V_2)\,dt\right]$$

$$V_O = \frac{V_2(T_2 - T_1)}{T} \tag{2.13}$$

Equation (2.12) in (2.13) gives

$$V_O = \frac{V_1V_2}{V_{SAT}} \tag{2.14}$$

2.4 TIME DIVISION MULTIPLIER NO REFERENCE—TYPE II

The multipliers using the time division principle without using any reference clock is shown in Figure 2.7, and its associated waveforms are shown in Figure 2.8.

Initially the 555 timer output is HIGH. The op-amp OA_3 gives $-V_1$ to the inverting terminal of differential integrator (the switch S_1 is closed, and the op-amp OA_3 will work as a non-inverting amplifier). The output of differential integrator will be

$$V_{T1} = \frac{1}{R_1C_1}\int (V_O + V_1)dt$$

$$V_{T1} = \frac{(V_O + V_1)}{R_1C_1} t \tag{2.15}$$

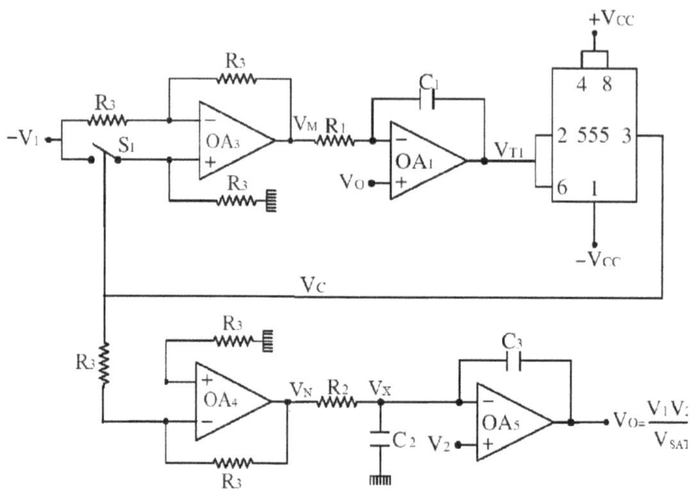

Figure 2.7 Time division multiplier without reference clock—II.

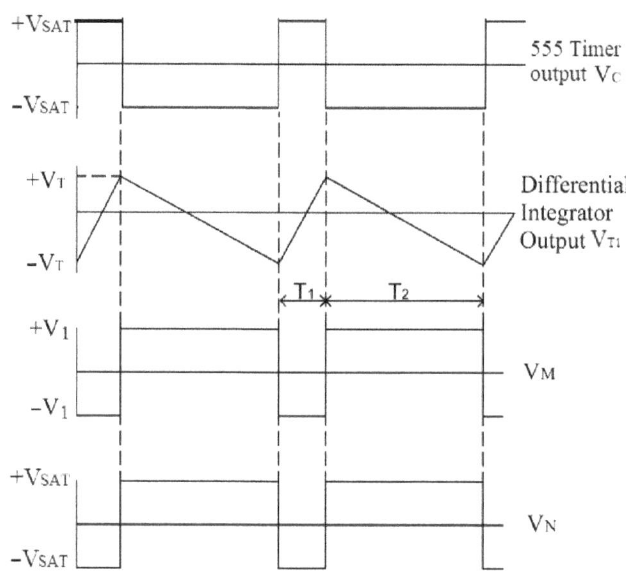

Figure 2.8 Associated waveforms of Figure 2.7.

The output of the differential integrator rises toward positive saturation, and when it reaches the voltage level of $+V_T$, the 555 timer output becomes LOW. The op-amp OA_3 gives $+V_1$ to the inverting terminal of the differential integrator (the switch S_1 is opened, and the op-amp OA_3 will work as an inverting amplifier). Now the output of the differential integrator will be

$$V_{T1} = \frac{1}{R_1C_1}\int (V_O - V_1)dt$$

$$V_{T1} = -\frac{(V_1 - V_O)}{R_1C_1}t \tag{2.16}$$

The output of the differential integrator reverses toward negative saturation, and when it reaches the voltage level $-V_T$, the 555 timer output becomes HIGH, and the cycle therefore repeats, to give an asymmetrical rectangular wave V_C at the output of the 555 timer.

$$V_T = \frac{V_{CC}}{3} \tag{2.17}$$

From the waveforms shown in Figure 2.8, it is observed that

$$T_1 = \frac{V_1 - V_O}{2V_1}T,\ T_2 = \frac{V_1 + V_O}{2V_1}T,\ T = T_1 + T_2 \tag{2.18}$$

Another rectangular wave V_N, with $\pm V_{SAT}$ as the peak to peak value, is generated at the output of the inverting amplifier OA_4. The R_2C_2 low pass filter gives the average value of this pulse train V_N and is given as

$$V_X = \frac{1}{T}\left[\int_O^{T_2} V_{SAT}\,dt + \int_{T_2}^{T_1+T_2} (-V_{SAT})dt\right]$$

$$V_X = \frac{V_{SAT}(T_2 - T_1)}{T} \tag{2.19}$$

Equation (2.17) in (2.19) gives

$$V_X = \frac{V_O V_{SAT}}{V_1} \tag{2.20}$$

The op-amp OA_5 is at negative closed loop configuration, and a positive dc voltage is ensured in the feedback loop. Hence its non-inverting terminal voltage is equal to its inverting terminal voltage, i.e.,

$$V_2 = V_X \tag{2.21}$$

From equations (2.20) and (2.21),

$$V_O = \frac{V_1 V_2}{V_{SAT}} \tag{2.22}$$

2.5 MULTIPLIER FROM 555 ASTABLE MULTIVIBRATOR

2.5.1 Multiplier from 555 Astable Multivibrator—Type I

The circuit diagram of a multiplier using the 555 timer astable multivibrator is shown in Figure 2.9, and its associated waveforms are shown in Figure 2.10. Refer to the internal diagram of the 555 timer IC shown in Figure 0.1. Initially when we switch on the power supply, the output of the upper comparator CMP_1 will be LOW, i.e., R = 0, and the output of the lower comparator CMP_2 will be HIGH, i.e., S = 1. The flip flop outputs are Q = 1 and Q′ = 0. The timer output at pin 3 will be HIGH, transistor Q_1 is OFF, and hence the discharge pin 7 is at the open position.

The capacitor C_1 is charging toward $+V_{CC}$ through the resistors R_1 and R_2 with a time constant of $(R_1+R_2)C_1$, and its voltage is rising exponentially. When the capacitor voltage is rising above the voltage V_1, the output of the

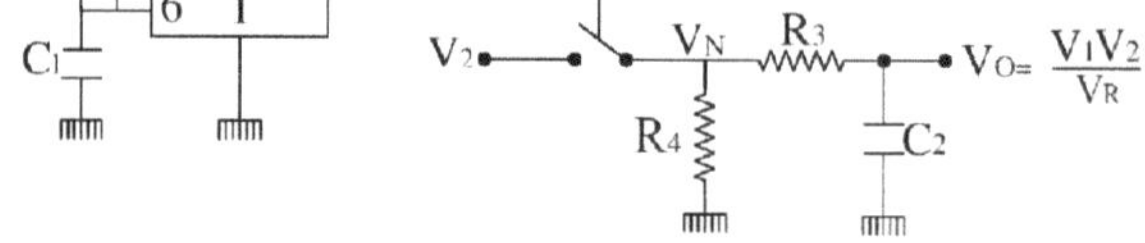

Figure 2.9 Multiplier with 555 timer astable multivibrator—type I.

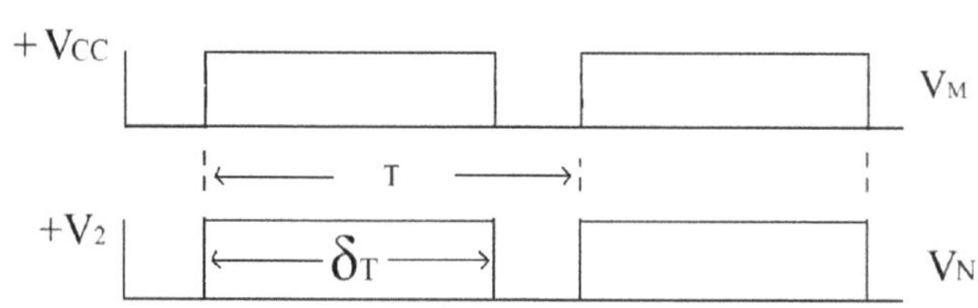

Figure 2.10 Associated waveforms of Figure 2.9.

upper comparator CMP_1 becomes HIGH, i.e., R = 1, and the output of the lower comparator CMP_2 becomes LOW, i.e., S = 0. The flip flop outputs are Q = 0 and Q′ = 1. The timer output at pin 3 will be LOW, transistor Q_1 is ON, and hence the discharge pin 7 is at GND potential. Now the capacitor C_1 is discharging to GND potential through the resistor R_2 with a time constant of R_2C. When the capacitor voltage falls below 1/3 V_{CC}, the output of the upper comparator CMP_1 becomes LOW, i.e., R = 0, and the output of the lower comparator CMP_2 becomes HIGH, i.e., S = 1. The flip flop outputs are Q = 1 and Q′ = 0. The timer output at pin 3 will be HIGH, transistor Q_1 is OFF, and hence the discharge pin 7 is at the open position.

Now the capacitor starts charging toward $+V_{CC}$, and the cycle therefore repeats to produce periodic pulses at the output pin 3 of the 555 timer.

The ON time of the 555 timer output V_M is proportional to V_1, which is applied at its pin 5. During the ON time δ_T, the second input voltage V_2 is connected to R_3C_2 low pass filter (switch S_1 is closed). During the OFF time of V_M, zero voltage exists on the R_3C_2 low pass filter (switch S_1 is opened). Another rectangular waveform V_N, with V_2 as the peak value, is generated at the output of switch S_1.

$$\delta_T = \frac{V_1}{V_R} T \tag{2.23}$$

The R_3C_2 low pass filter gives the average value of this pulse train V_N and is given as

$$V_O = \frac{1}{T}\int_0^{\delta_T} V_2 dt = \frac{V_2}{T}\delta_T$$

$$V_O = \frac{V_1 V_2}{V_R} \tag{2.24}$$

where V_R is a constant value.

2.5.2 Multiplier from 555 Astable Multivibrator—Type II

The circuit diagram of a divider using the 555 astable multivibrator is shown in Figure 2.11, and its associated waveforms are shown in Figure 2.12. Refer to the internal diagram of the 555 timer IC shown in Figure 0.1. Initially when we switch on the power supply, the output of the upper comparator CMP_1 will be LOW, i.e., R = 0, and the output of the lower comparator CMP_2 will be HIGH, i.e., S = 1. The flip flop outputs are Q = 1 and Q′ = 0. The timer output at pin 3 will be HIGH, transistor Q_1 is OFF, and hence the discharge pin 7 is at the open position.

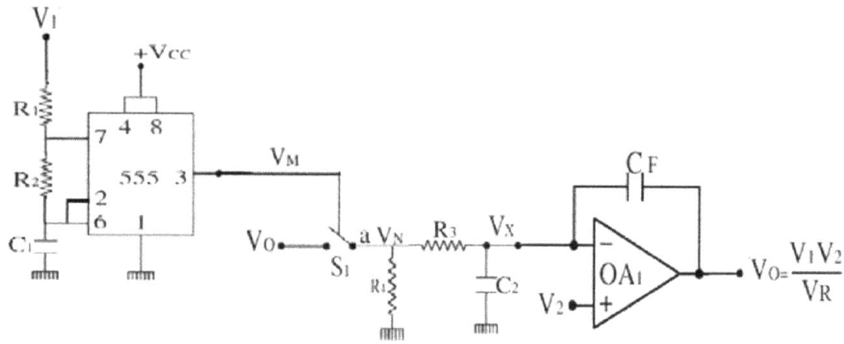

Figure 2.11 Multiplier from 555 astable—type II.

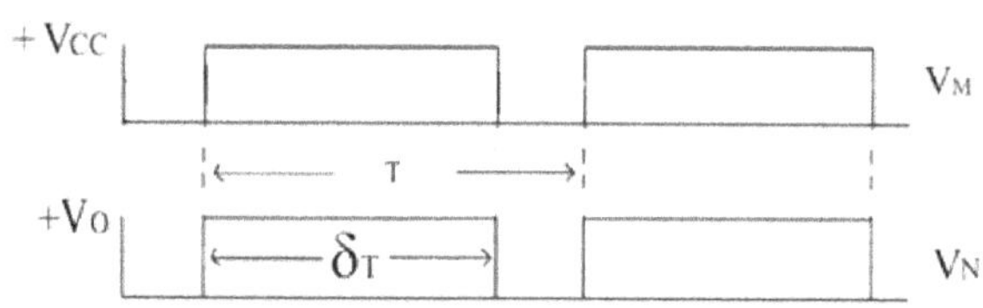

Figure 2.12 Associated waveforms of Figure 2.11.

The capacitor C_1 is charging toward V_1 through the resistors R_1 and R_2 with a time constant of $(R_1+R_2)C_1$, and its voltage is rising exponentially. When the capacitor voltage rises above the voltage 2/3 V_{CC}, the output of the upper comparator CMP_1 becomes HIGH, i.e., R = 1, and the output of the lower comparator CMP_2 becomes LOW, i.e., S = 0. The flip flop outputs are Q = 0 and Q′ = 1. The timer output at pin 3 will be LOW, transistor Q_1 is ON, and hence the discharge pin 7 is at GND potential. Now the capacitor C_1 is discharging to GND potential through the resistor R_2 with a time constant of R_2C. When the capacitor voltage falls below 1/3 V_{CC}, the output of the upper comparator CMP_1 becomes LOW, i.e., R = 0, and the output of the lower comparator CMP_2 becomes HIGH, i.e., S = 1. The flip flop outputs are Q = 1 and Q′ = 0. The timer output at pin 3 will be HIGH, transistor Q_1 is OFF, and hence the discharge pin 7 is at the open position.

Now the capacitor starts charging toward V_1, and the cycle therefore repeats to produce periodic pulses at the output pin 3 of the 555 timer.

The ON time of the 555 timer output V_M is inversely proportional to V_1. During the ON time δ_T, the voltage V_O is connected to the R_3C_2 low pass filter (switch S_1 is closed). During the OFF time of V_M, zero voltage exists on the R_3C_2 low pass filter (switch S_1 is opened). Another rectangular waveform

V_N, with V_O as the peak value, is generated at the output of switch S_1. The ON time δ_T of this rectangular pulse V_N is given as

$$\delta_T = \frac{V_R}{V_1} T \tag{2.25}$$

where V_R is a constant value.

The R_3C_2 low pass filter gives the average value of this pulse train V_N and is given as

$$V_X = \frac{1}{T}\int_0^{\delta_T} V_O dt = \frac{V_O}{T}\delta_T$$

$$V_X = \frac{V_O}{V_1} V_R \tag{2.26}$$

The op-amp OA_1 is at a negative closed feedback configuration, and a positive dc voltage is ensured in the feedback loop. Hence its non-inverting terminal voltage must be equal to its inverting terminal voltage.

$$V_2 = V_X \tag{2.27}$$

From equations (2.26) and (2.27),

$$V_O = \frac{V_1 V_2}{V_R} \tag{2.28}$$

2.6 MULTIPLIER FROM 555 MONOSTABLE MULTIVIBRATOR

2.6.1 Type I

The circuit diagram of a multiplier using the 555 timer monostable multivibrator is shown in Figure 2.13, and its associated waveforms are shown in Figure 2.14. Refer to the internal diagram of the 555 timer IC shown in Figure 0.1. Initially when the power supply is switched on, the output of the upper comparator CMP_1 will be LOW, i.e., R = 0, and the output of the lower comparator CMP_2 will be HIGH, i.e., S = 1. The flip flop outputs are Q = 1 and Q′ = 0. The timer output at pin 3 will be HIGH, transistor Q_1 is OFF, and hence the discharge pin 7 is at the open position. The capacitor C_1 is charging toward $+V_{CC}$ through the resistor R_1. The capacitor voltage is rising exponentially, and when it reaches the value of V_1, the output of the upper comparator CMP_1 becomes HIGH, i.e., R = 1, and the output of the lower comparator CMP_2 becomes LOW, i.e., S = 0. The flip flop outputs are

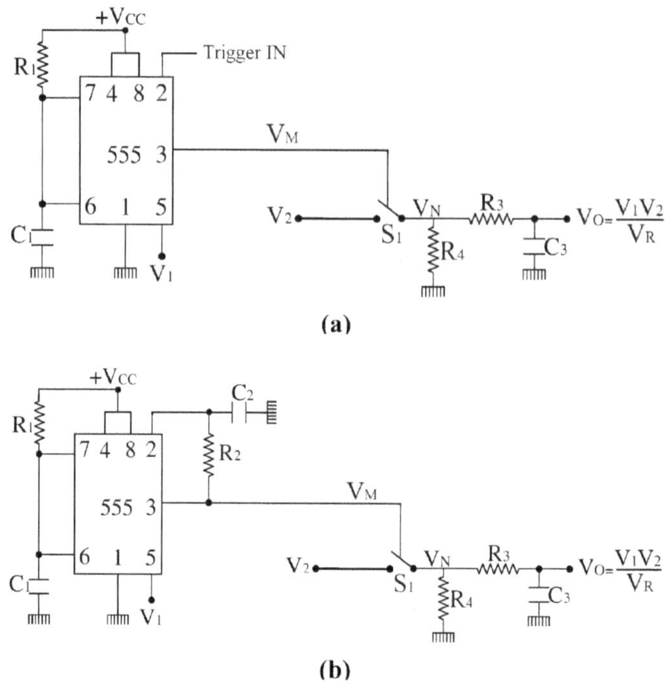

Figure 2.13 (a) Multiplier from 555 monostable. (b) Multiplier with 555 re-trigger monostable multivibrator.

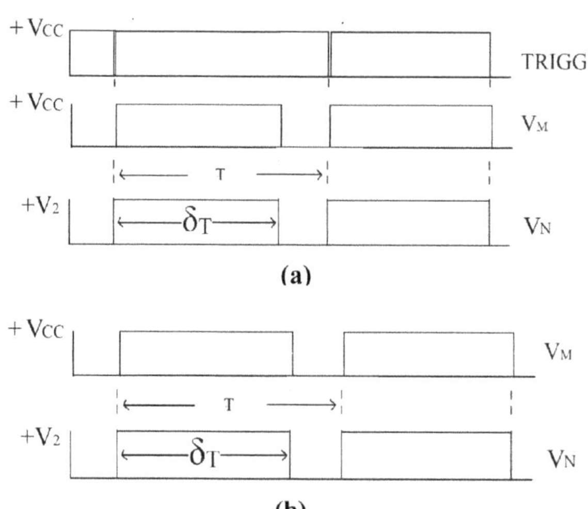

Figure 2.14 (a) Associated waveforms of Figure 2.13(a). (b) Associated waveforms of Figure 2.13(b).

Q = 0 and Q′ = 1. The timer output at pin 3 will be LOW, transistor Q_1 is ON, and hence the discharge pin 7 is at GND potential. Now the capacitor C_1 is short circuited, zero voltage exists at pin 6, and the output of the upper comparator CMP_1 becomes LOW, i.e., R = 0. A trigger pulse is applied at pin 2, and when the trigger voltage comes down to 1/3 V_{CC}, the output of the lower comparator CMP_2 becomes HIGH, i.e., S = 1. The flip flop outputs are Q = 1 and Q′ = 0. The timer output at pin 3 will be HIGH, transistor Q_1 is OFF, and hence the discharge pin 7 is at the open position.

Now the capacitor C_1 is charging toward $+V_{CC}$, and the sequence therefore repeats for every trigger input pulse.

The ON time of the 555 timer output V_M is proportional to V_1, which is applied at its pin 5. The 555 timer output controls switch S_1. During the ON time δ_T, the second input voltage V_2 is connected to the R_3C_3 low pass filter (switch S_1 is closed). During the OFF time of V_M, zero voltage is connected to the R_3C_3 low pass filter (switch S_1 is opened). Another rectangular waveform V_N, with V_2 as the peak value, is generated at the output of switch S_1. The ON time δ_T of this rectangular waveform V_N is given as

$$\delta_T = \frac{V_1}{V_R} T \tag{2.29}$$

The R_3C_3 low pass filter gives the average value of this pulse train V_N and is given as

$$V_O = \frac{1}{T} \int_0^{\delta_T} V_2 dt = \frac{V_2}{T} \delta_T$$

$$V_O = \frac{V_1 V_2}{V_R} \tag{2.30}$$

where V_R is a constant value.

A multiplier using a re-trigger monostable multivibrator is shown in Figure 2.13(b).

2.6.2 Multiplier from 555 Monostable Multivibrator—Type II

The circuit diagram of a multiplier using the 555 monostable multivibrator is shown in Figure 2.15, and its associated waveforms are shown in Figure 2.16. Refer to the internal diagram of the 555 timer IC shown in Figure 0.1. Initially when the power supply is switched on, the output of the upper comparator CMP_1 will be LOW, i.e., R = 0, and the output of the lower comparator CMP_2 will be HIGH, i.e., S = 1. The flip flop outputs are Q = 1 and Q′ = 0. The timer output at pin 3 will be HIGH, transistor Q_1 is

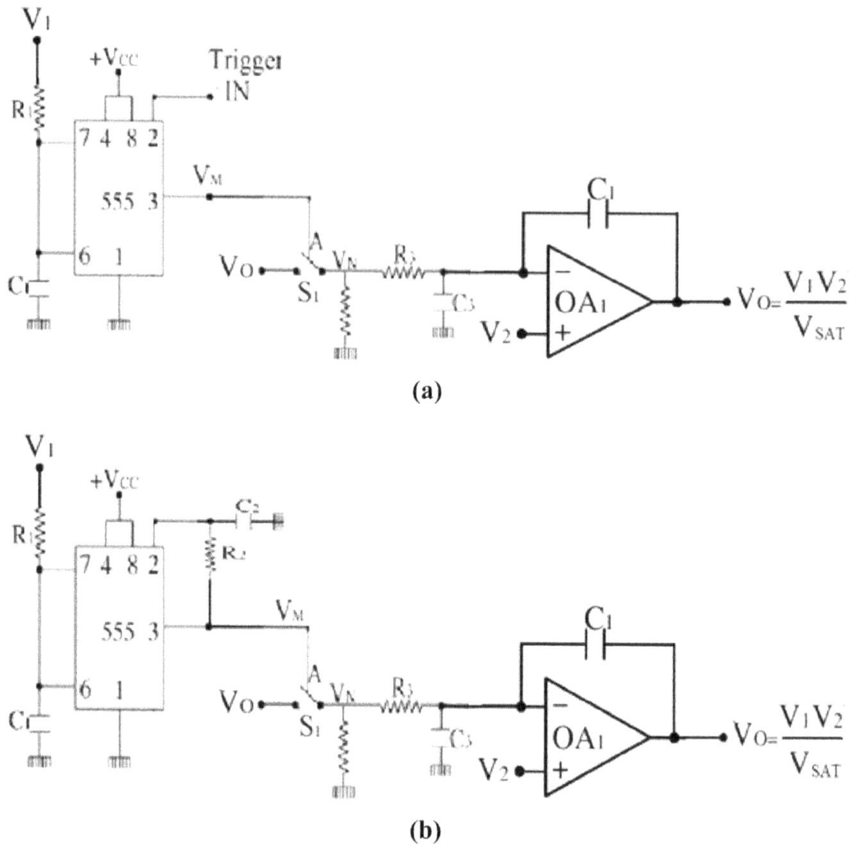

Figure 2.15 (a) Multiplier using 555 timer monostable multivibrator. (b) Multiplier using re-trigger monostable multivibrator.

OFF, and hence the discharge pin 7 is at the open position. The capacitor C_1 is charging toward V_1 through the resistor R_1. The capacitor voltage is rising exponentially, and when it reaches the value of 2/3 V_{CC}, the output of the upper comparator CMP_1 becomes HIGH, i.e., R = 1, and the output of the lower comparator CMP_2 becomes LOW, i.e., S = 0. The flip flop outputs are Q = 0 and Q′ = 1. The timer output at pin 3 will be LOW, transistor Q_1 is ON, and hence the discharge pin 7 is at GND potential. Now the capacitor C_1 is short circuited, zero voltage exists at pin 6, the output of the upper comparator CMP_1 becomes LOW, i.e., R = 0. A trigger pulse is applied at pin 2, and when the trigger voltage comes down to 1/3 V_{CC}, the output of the lower comparator CMP_2 becomes HIGH, i.e., S = 1. The flip flop outputs are Q = 1 and Q′ = 0. The timer output at pin 3 will be HIGH, transistor Q_1 is OFF, and hence the discharge pin 7 is at the open position.

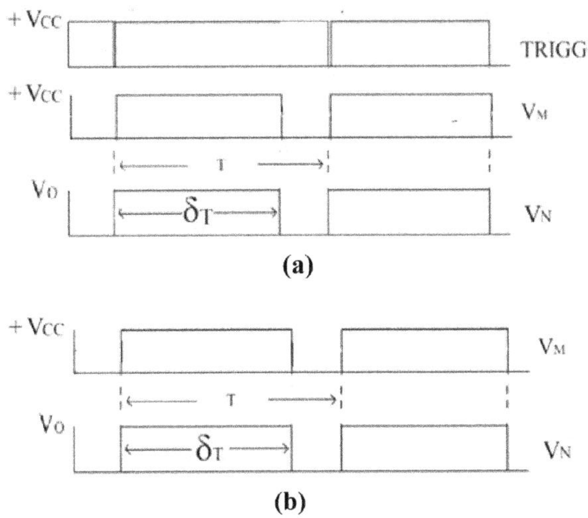

Figure 2.16 (a) Associated waveforms of Figure 2.15(a). (b) Associated waveforms of Figure 2.15(b).

Now the capacitor C_1 is charging toward V_1, and the sequence therefore repeats for every trigger input pulse.

The ON time of the 555 timer output V_M is inversely proportional to V_1. The output of the 555 timer control switch S_1. During the ON time δ_T, the voltage V_O is connected to the R_3C_3 low pass filter (switch S_1 is closed). During the OFF time of V_M, zero voltage exists on the R_3C_3 low pass filter (switch S_1 is opened). Another rectangular waveform V_N, with V_O as the peak value, is generated at the output of switch S_1.

$$\delta_T = \frac{V_R}{V_1} T \tag{2.31}$$

The R_3C_3 low pass filter gives the average value of this pulse train V_N and is given as

$$V_X = \frac{1}{T}\int_0^{\delta_T} V_O dt = \frac{V_O}{T}\delta_T$$

$$V_X = \frac{V_O}{V_1} V_R \tag{2.32}$$

where V_R is a constant value.

The op-amp OA_1 is at a negative closed feedback configuration, and a positive dc voltage is ensured in the feedback loop. Hence its non-inverting terminal voltage must be equal to its inverting terminal voltage.

$$V_2 = V_X \tag{2.33}$$

From equations (2.32) and (2.33),

$$V_O = \frac{V_1 V_2}{V_R} \tag{2.34}$$

Figure 2.15(b) shows the multiplier using a re-trigger monostable multivibrator.

Chapter 3

Time Division Dividers (TDD)—Multiplexing

3.1 SAW TOOTH WAVE BASED TIME DIVISION DIVIDERS

The circuit diagrams of saw tooth wave based time division dividers are shown in Figure 3.1, and their associated waveforms are shown in Figure 3.2. A saw tooth wave V_{S1} of peak value V_R and during time period T is generated by the 555 timer IC.

The circuit working operation of a saw tooth wave generator is given in chapter 1.

In the circuits of Figure 3.1, the comparator OA_2 compares the saw tooth wave V_{S1} of peak value V_R with the input voltage V_1 and produces a rectangular waveform V_M at its output. The ON time δ_T of this rectangular waveform V_M is given as

$$\delta_T = \frac{V_1}{V_R} T \tag{3.1}$$

The rectangular pulse V_M controls the multiplexer M_1. When V_M is HIGH, another input voltage V_O is connected to the R_3C_2 low pass filter ('ay' is connected to 'a'). When V_M is LOW, zero voltage is connected to the R_3C_2 low pass filter ('ax' is connected to 'a'). Another rectangular pulse V_N with maximum value of V_O is generated at the multiplexer M_1 output. The R_3C_2 low pass filter gives the average value of this pulse train V_N and is given as

$$V_X = \frac{1}{T}\int_0^{\delta_T} V_O dt \tag{3.2}$$

$$V_X = \frac{V_O}{T}\delta_T \tag{3.3}$$

Equation (3.1) in (3.3) gives

DOI: 10.1201/9781003362968-3

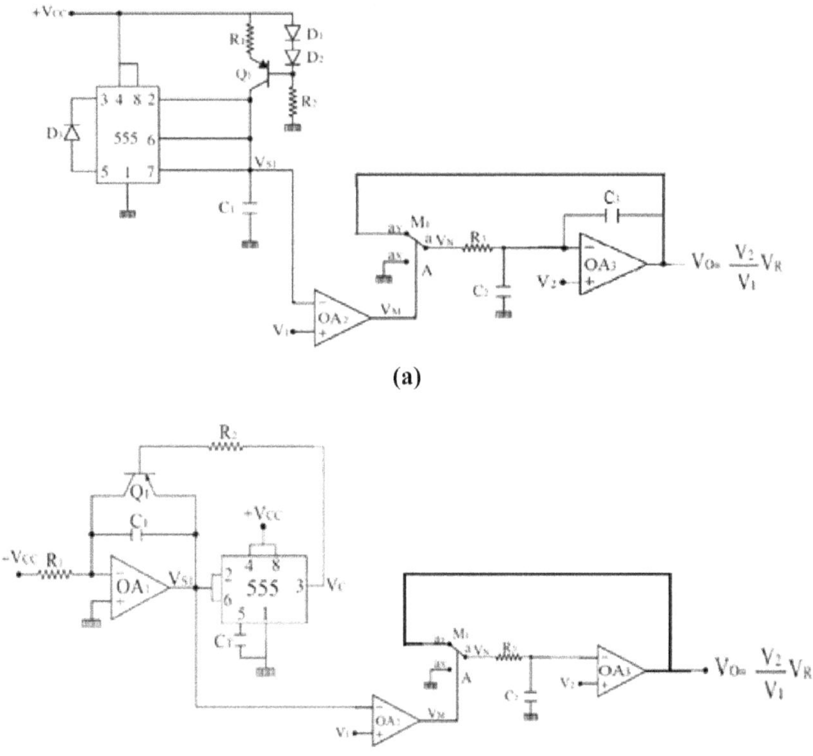

(a)

(b)

Figure 3.1 (a) Saw tooth wave based time division divider—type I. (b) Saw tooth wave based time division divider—type II.

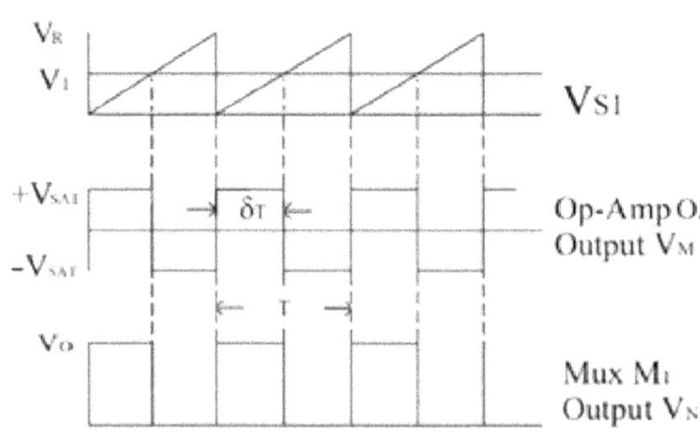

Figure 3.2 Associated waveforms of Figure 3.1.

$$V_X = \frac{V_1 V_O}{V_R} \tag{3.4}$$

where $V_R = 2/3\ V_{CC}$.

The op-amp OA_3 is kept in a negative closed loop configuration, and a positive dc voltage is ensured in the feedback. Hence its inverting terminal voltage will be equal to its non-inverting terminal voltage, i.e.,

$$V_X = V_2 \tag{3.5}$$

From equations (3.4) and (3.5)

$$V_O = \frac{V_2}{V_1} V_R \tag{3.6}$$

3.2 TRIANGULAR WAVE REFERENCED TIME DIVISION DIVIDERS

The circuit diagrams of triangular wave based dividers are shown in Figure 3.3, and their associated waveforms are shown in Figure 3.4. In Figure 3.3(a), a triangular wave V_{T1} with $\pm V_T$ peak to peak value and time period T is generated by the 555 timer. The working operation of a triangular wave generator is described in chapter 1.

In the circuits of Figure 3.3(a) and (b), the output voltage V_O is compared with the generated triangular wave V_{T1} by the comparator on OA_2. An asymmetrical rectangular waveform V_M is generated at the comparator OA_2 output. From the waveforms shown in Figure 3.5, it is observed that

$$T_1 = \frac{V_T - V_O}{2V_T} T,\ T_2 = \frac{V_T + V_O}{2V_T} T,\ T = T_1 + T_2 \tag{3.7}$$

This rectangular wave V_M is given as the control input to the multiplexer M_1. The multiplexer M_1 connects the input voltage $+V_1$ during T_2 ('ay' is connected to 'a') and $-V_1$ during T_1 ('ax' is connected to 'a'). Another rectangular asymmetrical wave V_N with a peak to peak value of $\pm V_1$ is generated at the multiplexer M_1 output. The R_3C_3 low pass filter gives the average value of the pulse train V_N, which is given as

$$V_X = \frac{1}{T}\left[\int_O^{T_2} V_1\, dt + \int_{T_2}^{T_1+T_2} (-V_1)\, dt\right] = \frac{V_1}{T}(T_2 - T_1) \tag{3.8}$$

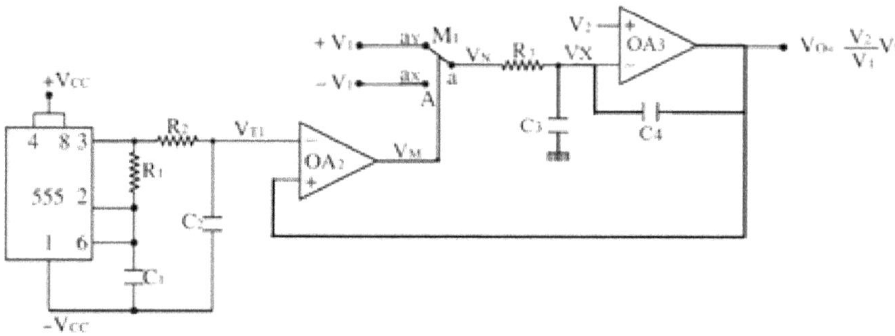

(a)

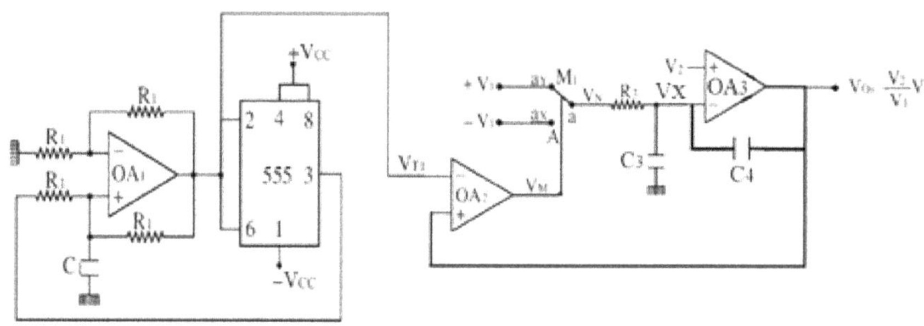

(b)

Figure 3.3 (a) Triangular wave based divider—type I. (b) Triangular wave based divider—type II.

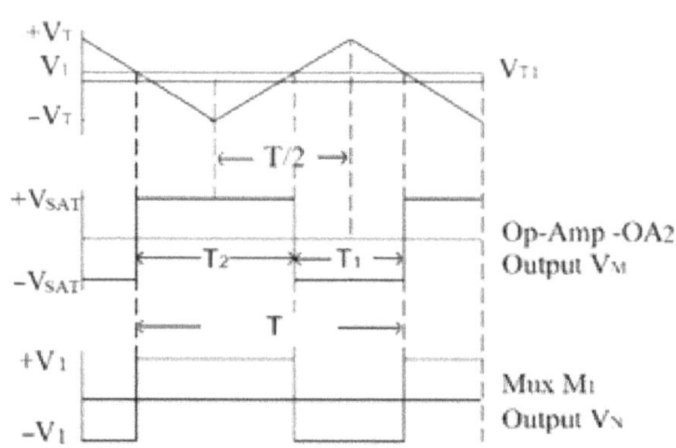

Figure 3.4 Associated waveforms of Figure 3.3(a) and (b).

Equation (3.7) in (3.8) gives

$$V_X = \frac{V_1 V_O}{V_T} \tag{3.9}$$

where $V_T = V_{CC}/3$. (3.10)

The op-amp OA_3 is kept in a negative closed loop configuration, and a positive dc voltage is ensured in the feedback. Hence its inverting terminal voltage will be equal to its non-inverting terminal voltage, i.e.,

$$V_X = V_2 \tag{3.11}$$

From equations (3.9) and (3.11),

$$V_O = \frac{V_2}{V_1} V_T \tag{3.12}$$

3.3 TIME DIVISION DIVIDER WITH NO REFERENCE—TYPE I

The divider using the time division principle without using any reference clock is shown in Figure 3.5, and its associated waveforms are shown in Figure 3.6.

Initially the 555 timer output is HIGH. The multiplexer M_1 connects $-V_1$ to the differential integrator composed of resistor R_1, capacitor C_1, and op-amp OA_1 ('ay' is connected to 'a'). The output of differential integrator will be

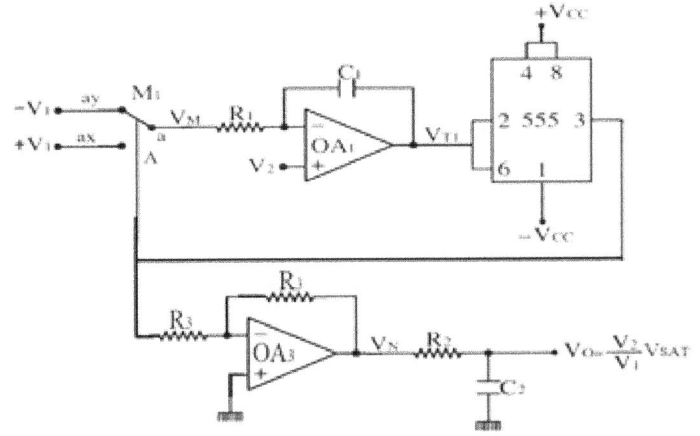

Figure 3.5 Time division divider without reference clock.

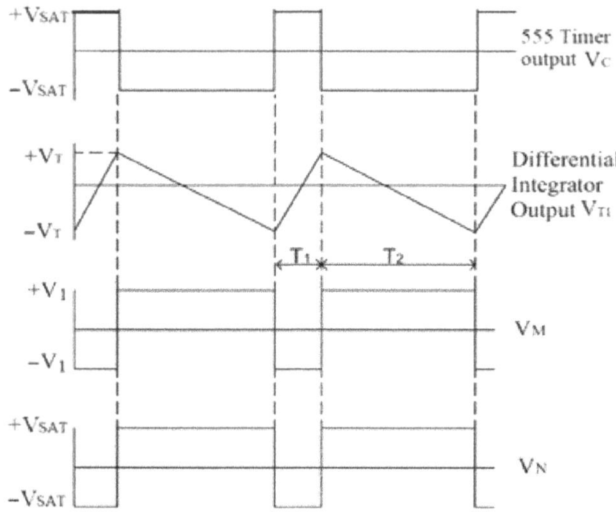

Figure 3.6 Associated waveforms of Figure 3.5.

$$V_{T1} = \frac{1}{R_1C_1}\int (V_2 + V_1)dt$$

$$V_{T1} = \frac{(V_1 + V_2)}{R_1C_1}t \tag{3.13}$$

The output of the differential integrator rises toward positive saturation, and when it reaches the voltage level of $+V_T$, the 555 timer output becomes LOW. The multiplexer M_1 connects $+V_1$ to the differential integrator composed by resistor R_1, capacitor C_1, and op-amp OA_1 ('ax' is connected to 'a'). Now the output of the differential integrator will be

$$V_{T1} = \frac{1}{R_1C_1}\int (V_2 - V_1)dt$$

$$V_{T1} = -\frac{(V_1 - V_2)}{R_1C_1}t \tag{3.14}$$

The output of the differential integrator reverses toward negative saturation, and when it reaches the voltage level $-V_T$, the 555 timer output becomes HIGH, and the cycle therefore repeats, to give an asymmetrical rectangular wave V_C at the output of the 555 timer.

$$V_T = \frac{V_{CC}}{3} \tag{3.15}$$

From the waveforms shown in Figure 3.6, it is observed that

$$T_1 = \frac{V_{1-}V_2}{2V_1}T, T_2 = \frac{V_1 + V_2}{2V_1}T, T = T_1 + T_2 \tag{3.16}$$

Another rectangular wave V_N is generated at the output of the inverting amplifier OA_3. The R_2C_2 low pass filter gives the average value of this pulse train V_N and is given as

$$V_O = \frac{1}{T}\left[\int_O^{T_2} V_{SAT}\,dt + \int_{T_2}^{T_1+T_2} (-V_{SAT})dt\right]$$

$$V_O = \frac{V_{SAT}(T_2 - T_1)}{T} \tag{3.17}$$

Equation (3.16) in (3.17) gives

$$V_O = \frac{V_2}{V_1}V_{SAT} \tag{3.18}$$

3.4 TIME DIVISION DIVIDER WITH NO REFERENCE—TYPE II

The divider using the time division principle without using any reference clock is shown in Figure 3.7, and its associated waveforms are shown in Figure 3.8.

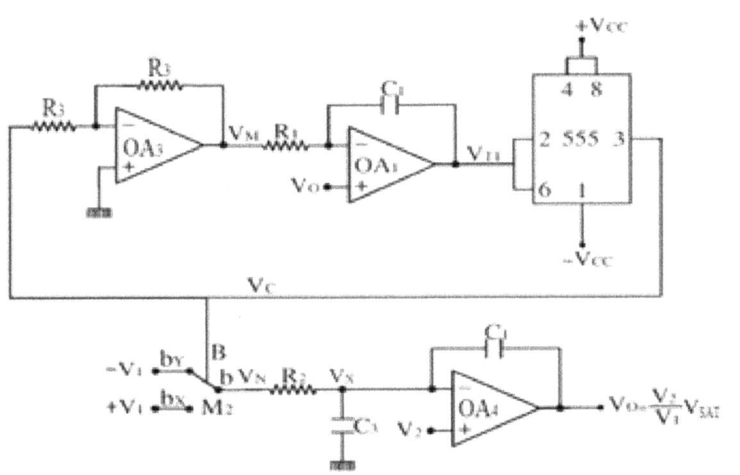

Figure 3.7 Divider without reference clock—Type II.

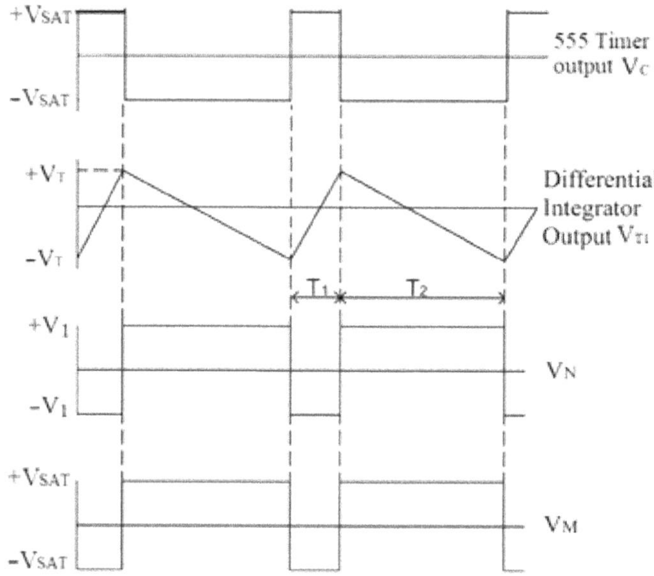

Figure 3.8 Associated waveforms of Figure 3.7.

Initially the 555 timer output is HIGH. The inverting amplifier OA_3 output will be $-V_{SAT}$. The output of differential integrator will be

$$V_{T1} = \frac{1}{R_1C_1}\int (V_O + V_{SAT})dt$$

$$V_{T1} = \frac{(V_O + V_{SAT})}{R_1C_1}t \tag{3.19}$$

The output of the differential integrator rises toward positive saturation, and when it reaches the voltage level of $+V_T$, the 555 timer output becomes LOW. The inverting amplifier OA_3 output will be $+V_{SAT}$. Now the output of differential integrator will be

$$V_{T1} = \frac{1}{R_1C_1}\int (V_O - V_{SAT})dt$$

$$V_{T1} = -\frac{(V_{SAT} - V_O)}{R_1C_1}t \tag{3.20}$$

The output of differential integrator reverses toward negative saturation, and when it reaches the voltage level $-V_T$, the 555 timer output becomes HIGH, and the cycle therefore repeats to give an asymmetrical rectangular wave V_C at the output of the 555 timer.

$$V_T = \frac{V_{CC}}{3} \tag{3.21}$$

From the waveforms shown in Figure 3.8, it is observed that

$$T_1 = \frac{V_{SAT} - V_O}{2V_{SAT}} T, T_2 = \frac{V_{SAT} + V_O}{2V_{SAT}} T, T = T_1 + T_2 \tag{3.22}$$

The asymmetrical rectangular wave V_C controls the multiplexer M_2. The multiplexer M_2 connects $-V_1$ during the ON time T_2 ('by' is connected to 'b') and $+V_1$ during the OFF time T_1 of the rectangular wave V_C ('bx' is connected to 'b'). Another rectangular wave V_N with $\pm V_1$ as the peak to peak value is generated at the multiplexer M_2 output. The R_2C_2 low pass filter gives the average value of this pulse train V_N and is given as

$$V_X = \frac{1}{T}\left[\int_O^{T_2} V_1\,dt + \int_{T_2}^{T_1+T_2} (-V_1)dt\right]$$

$$V_X = \frac{V_1(T_2 - T_1)}{T} \tag{3.23}$$

Equations (3.22) in (3.23) gives

$$V_X = \frac{V_O V_1}{V_{SAT}} \tag{3.24}$$

The op-amp OA_4 is at a negative closed loop configuration, and a positive dc voltage is ensured in the feedback loop. Hence its non-inverting terminal voltage is equal to its inverting terminal voltage, i.e.,

$$V_2 = V_X \tag{3.25}$$

From equations (3.24) and (3.25),

$$V_O = \frac{V_2}{V_1} V_{SAT} \tag{3.26}$$

3.5 DIVIDER FROM 555 ASTABLE MULTIVIBRATOR

3.5.1 Type I

The circuit diagram of a divider using the 555 astable multivibrator is shown in Figure 3.9, and its associated waveforms are shown in Figure 3.10. Refer to the internal diagram of the 555 timer IC shown in Figure 0.1. Initially when we switch on the power supply, the output of the upper comparator CMP_1 will be LOW, i.e., R = 0, and the output of the lower comparator CMP_2 will be HIGH, i.e., S = 1. The flip flop outputs are Q = 1 and Q′ = 0. The timer output at pin 3 will be HIGH, transistor Q_1 is OFF, and hence the discharge pin 7 is at the open position.

The capacitor C_1 is charging toward V_1 through the resistors R_1 and R_2 with a time constant of $(R_1+R_2)C_1$, and its voltage is rising exponentially. When the capacitor voltage is rising above the voltage 2/3 V_{CC}, the output of the upper comparator CMP_1 becomes HIGH, i.e., R = 1, and the output of the lower comparator CMP_2 becomes LOW, i.e., S = 0. The flip flop outputs are Q = 0 and Q′ = 1. The timer output at pin 3 will be LOW, transistor Q_1

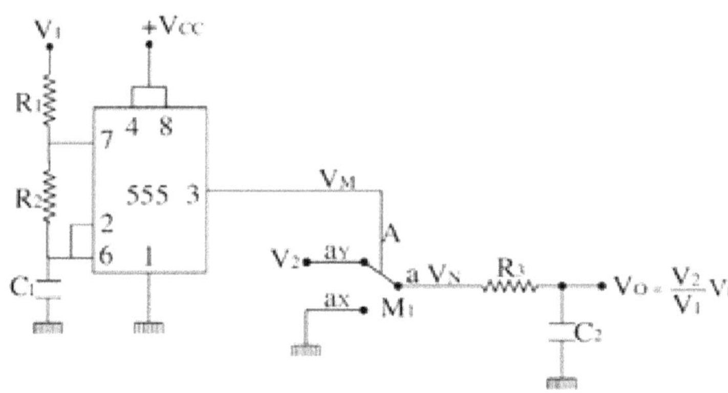

Figure 3.9 Divider from 555 astable.

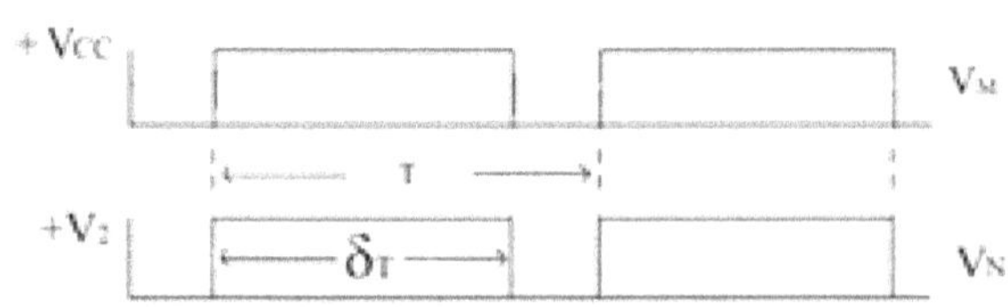

Figure 3.10 Associated waveforms of Figure 3.9.

is ON, and hence the discharge pin 7 is at GND potential. Now the capacitor C_1 is discharging to GND potential through the resistor R_2 with a time constant of R_2C. When the capacitor voltage falls below 1/3 V_{CC}, the output of the upper comparator CMP_1 becomes LOW, i.e., R = 0, and the output of the lower comparator CMP_2 becomes HIGH, i.e., S = 1. The flip flop outputs are Q = 1 and Q′ = 0. The timer output at pin 3 will be HIGH, transistor Q_1 is OFF, and hence the discharge pin 7 is at the open position.

Now the capacitor starts charging toward V_1, and the cycle therefore repeats to produce periodic pulses at the output pin 3 of the 555 timer.

The ON time of the 555 timer output V_M is inversely proportional to V_1. During the ON time δ_T, the second input voltage V_2 is connected to the R_3C_2 low pass filter ('ay' is connected to 'a'). During the OFF time of V_M, zero voltage is connected to the R_3C_2 low pass filter ('ax' is connected to 'a'). Another rectangular waveform V_N with V_2 as peak value is generated at the output of multiplexer M_1. The ON time δ_T of this rectangular pulse V_N is given as

$$\delta_T = \frac{V_R}{V_1} T \tag{3.27}$$

where V_R is a constant value.

The R_3C_2 low pass filter gives the average value of this pulse train V_N and is given as

$$V_O = \frac{1}{T}\int_0^{\delta_T} V_2 dt = \frac{V_2}{T}\delta_T$$

$$V_O = \frac{V_2}{V_1} V_R \tag{3.28}$$

3.5.2 Divider from 555 Astable Multivibrator—Type II

The circuit diagram of a divider using the 555 timer astable multivibrator is shown in Figure 3.11, and its associated waveforms are shown in Figure 3.12. Refer to the internal diagram of the 555 timer IC shown in Figure 0.1. Initially when we switch on the power supply, the output of the upper comparator CMP_1 will be LOW, i.e., R = 0, the output of the lower comparator CMP_2 will be HIGH, i.e., S = 1. The flip flop outputs are Q = 1 and Q′ = 0. The timer output at pin 3 will be HIGH, transistor Q_1 is OFF, and hence the discharge pin 7 is at the open position.

The capacitor C_1 is charging toward $+V_{CC}$ through the resistors R_1 and R_2 with a time constant of $(R_1+R_2)C_1$, and its voltage is rising exponentially. When the capacitor voltage is rising above the voltage V_1, the output of the

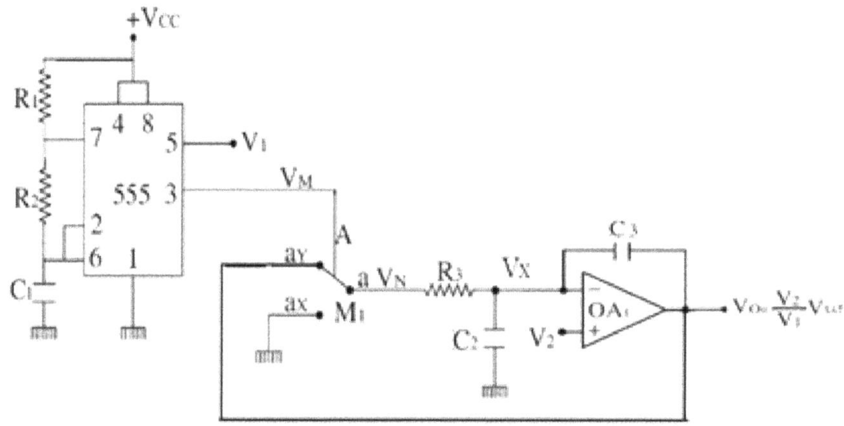

Figure 3.11 Divider with 555 timer astable multivibrator.

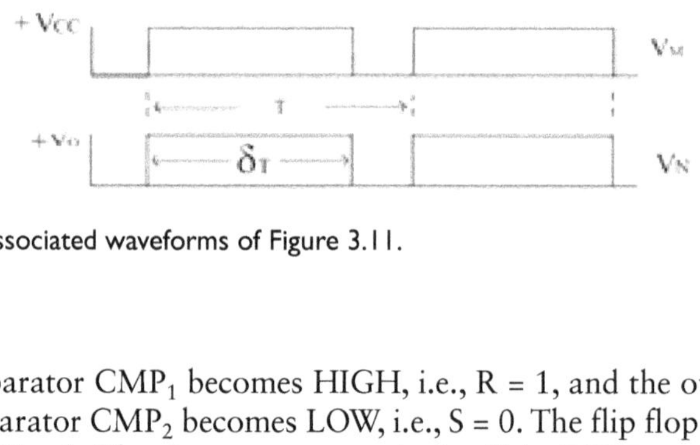

Figure 3.12 Associated waveforms of Figure 3.11.

upper comparator CMP_1 becomes HIGH, i.e., R = 1, and the output of the lower comparator CMP_2 becomes LOW, i.e., S = 0. The flip flop outputs are Q = 0 and Q′ = 1. The timer output at pin 3 will be LOW, transistor Q_1 is ON, and hence the discharge pin 7 is at GND potential. Now the capacitor C_1 is discharging to GND potential through the resistor R_2 with a time constant of R_2C. When the capacitor voltage falls below 1/3 V_{CC}, the output of the upper comparator CMP_1 becomes LOW, i.e., R = 0, and the output of the lower comparator CMP_2 becomes HIGH, i.e., S = 1. The flip flop outputs are Q = 1 and Q′ = 0. The timer output at pin 3 will be HIGH, transistor Q_1 is OFF, and hence the discharge pin 7 is at the open position.

Now the capacitor starts charging toward $+V_{CC}$, and the cycle therefore repeats to produce periodic pulses at the output pin 3 of the 555 timer.

The ON time of the 555 timer output V_M is proportional to V_1, which is applied at its pin 5. During the ON time δ_T, the output voltage V_O is connected to the R_3C_2 low pass filter ('ay' is connected to 'a'). During the OFF time of V_M, zero voltage is connected to the R_3C_2 low pass filter ('ax' is connected to 'a'). Another rectangular waveform V_N with V_2 as peak value is generated at the emitter of transistor.

$$\delta_T = \frac{V_1}{V_R} T \tag{3.29}$$

The R_3C_2 low pass filter gives the average value of this pulse train V_N and is given as

$$V_X = \frac{1}{T}\int_0^{\delta_T} V_O dt = \frac{V_O}{T}\delta_T$$

$$V_X = \frac{V_1 V_O}{V_R} \tag{3.30}$$

where V_R is a constant value.

The op-amp OA_1 is kept in negative closed loop configuration, and a positive dc voltage is ensured in the feedback. Hence its inverting terminal voltage will be equal to its non-inverting terminal voltage, i.e.,

$$V_X = V_2 \tag{3.31}$$

From equations (3.30) and (3.31),

$$V_O = \frac{V_2}{V_1} V_R \tag{3.32}$$

3.6 DIVIDER FROM 555 MONOSTABLE MULTIVIBRATOR

3.6.1 Type I

The circuit diagram of a divider using the 555 monostable multivibrator is shown in Figure 3.13(a), and its associated waveforms are shown in Figure 3.14(a). Refer to the internal diagram of the 555 timer IC shown in Figure 0.1. Initially when the power supply is switched on, the output of the upper comparator CMP_1 will be LOW, i.e., R = 0, and the output of the lower comparator CMP_2 will be HIGH, i.e., S = 1. The flip flop outputs are Q = 1 and Q' = 0. The timer output at pin 3 will be HIGH, transistor Q_1 is OFF, and hence the discharge pin 7 is at the open position. The capacitor C_1 is charging toward V_1 through the resistor R_1. The capacitor voltage rises exponentially, and when it reaches the value of 2/3 V_{CC}, the output of the upper comparator CMP_1 becomes HIGH, i.e., R = 1, and the output of the lower comparator CMP_2 becomes LOW, i.e., S = 0. The flip flop outputs are

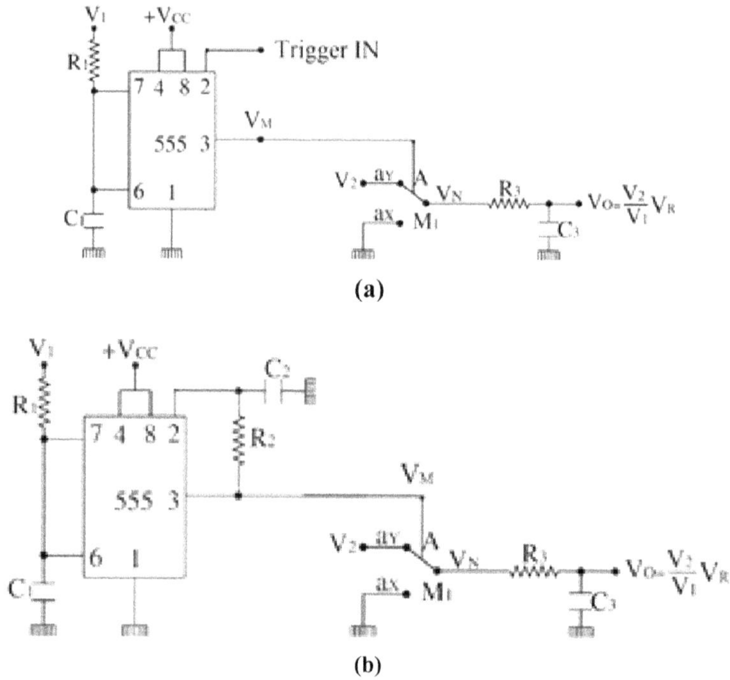

Figure 3.13 (a) Divider using 555 timer monostable multivibrator. (b) Divider using re-trigger monostable multivibrator.

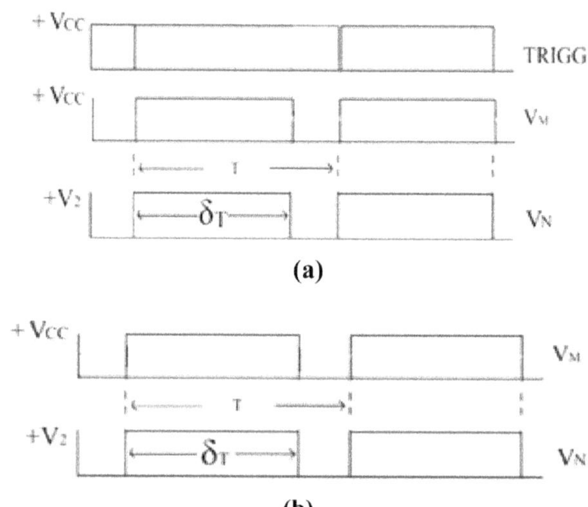

Figure 3.14 (a) Associated waveforms of Figure 3.13(a). (b) Associated waveforms of Figure 3.13(b).

$Q = 0$ and $Q' = 1$. The timer output at pin 3 will be LOW, transistor Q_1 is ON, and hence the discharge pin 7 is at GND potential. Now the capacitor C_1 is short circuited, zero volts exists at pin 6, and the output of the upper comparator CMP_1 becomes LOW, i.e., $R = 0$. A trigger pulse is applied at pin 2, and when the trigger voltage comes down to 1/3 V_{CC}, the output of the lower comparator CMP_2 becomes HIGH, i.e., $S = 1$. The flip flop outputs are $Q = 1$ and $Q' = 0$. The timer output at pin 3 will be HIGH, transistor Q_1 is OFF, and hence the discharge pin 7 is at the open position.

Now the capacitor C_1 is charging toward V_1, and the sequence therefore repeats for every trigger input pulse.

The ON time of the 555 timer output V_M is inversely proportional to V_1. The output of the 555 timer controls the multiplexer M_1. During the ON time δ_T, the second input voltage V_2 is connected to the R_3C_3 low pass filter ('ay' is connected to 'a'). During the OFF time of V_M, zero voltage is connected to the R_3C_3 low pass filter ('ax' is connected to 'a'). Another rectangular waveform V_N with V_2 as peak value is generated at the output of the multiplexer M_1.

$$\delta_T = \frac{V_R}{V_1} T \tag{3.33}$$

The R_3C_3 low pass filter gives the average value of this pulse train V_N and is given as

$$V_O = \frac{1}{T}\int_0^{\delta_T} V_2 dt = \frac{V_2}{T}\delta_T$$

$$V_O = \frac{V_2}{V_1} V_R \tag{3.34}$$

where V_R is a constant value.

The circuit diagram of a divider using the 555 re-trigger monostable multivibrator is shown in Figure 3.13(b), and its associated waveforms are shown in Figure 3.14(b).

3.6.2 Divider from 555 Monostable Multivibrator—Type II

The circuit diagram of a divider using the 555 timer monostable multivibrator is shown in Figure 3.15(a), and its associated waveforms are shown in Figure 3.16(a). Refer to the internal diagram of the 555 timer IC shown in Figure 0.1. Initially when the power supply is switched on, the output of the upper comparator CMP_1 will be LOW, i.e., $R = 0$, and the output of the lower comparator CMP_2 will be HIGH, i.e., $S = 1$. The flip flop outputs are $Q = 1$ and $Q' = 0$. The timer output at pin 3 will be

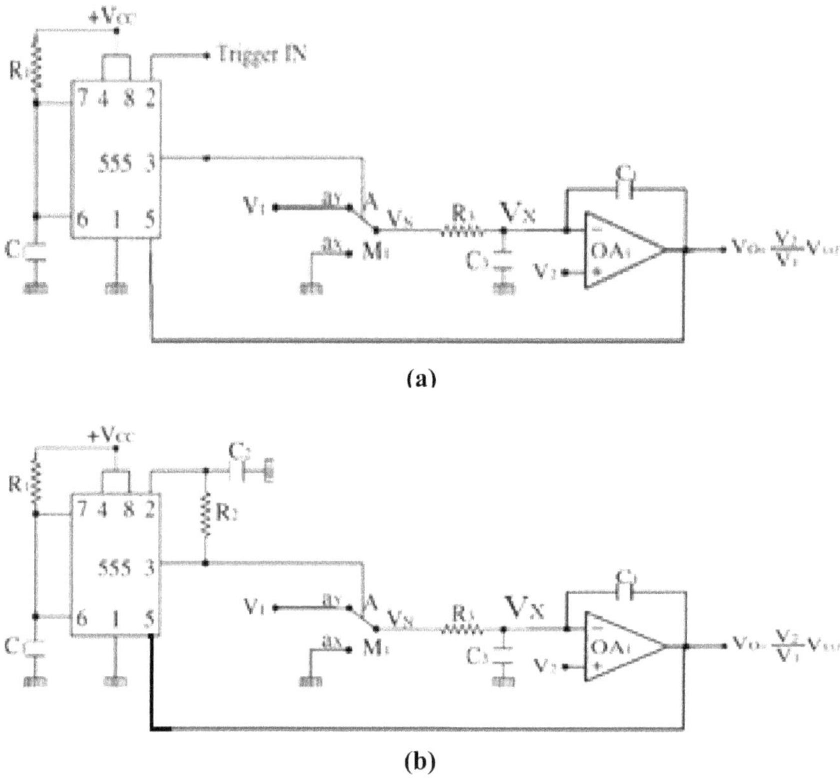

Figure 3.15 (a) Divider from 555 monostable. (b) Divider with 555 auto-trigger monostable multivibrator.

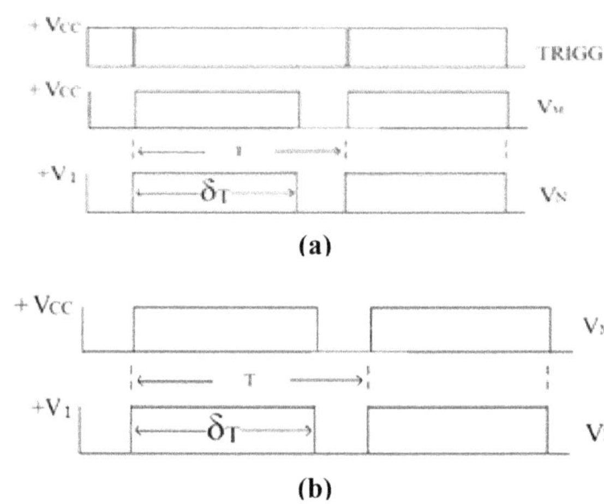

Figure 3.16 (a) Associated waveforms of Figure 3.16(a). (b) Associated waveforms of Figure 3.15(b).

HIGH, transistor Q_1 is OFF, and hence the discharge pin 7 is at the open position. The capacitor C_1 is charging toward $+V_{CC}$ through the resistor R_1. The capacitor voltage is rising exponentially, and when it reaches the value of V_O, the output of the upper comparator CMP_1 becomes HIGH, i.e., R = 1, and the output of the lower comparator CMP_2 becomes LOW, i.e., S = 0. The flip flop outputs are Q = 0 and Q′ = 1. The timer output at pin 3 will be LOW, transistor Q_1 is ON, and hence the discharge pin 7 is at GND potential. Now the capacitor C_1 is short circuited, zero voltage exists at pin 6, and the output of the upper comparator CMP_1 becomes LOW, i.e., R = 0. A trigger pulse is applied at pin 2, and when the trigger voltage comes down to 1/3 V_{CC}, the output of the lower comparator CMP_2 becomes HIGH, i.e., S = 1. The flip flop outputs are Q = 1 and Q′ = 0. The timer output at pin 3 will be HIGH, transistor Q_1 is OFF, and hence the discharge pin 7 is at the open position.

Now the capacitor C_1 is charging toward $+V_{CC}$, and the sequence therefore repeats for every trigger input pulse.

The ON time of the 555 timer output V_M is proportional to V_O, which is applied at its pin 5. The 555 timer output controls the multiplexer M_1. During the ON time δ_T, the input voltage V_1 is connected to R_3C_2 low pass filter ('ay' is connected to 'a'). During the OFF time of V_M, zero voltage is connected to the R_3C_2 low pass filter ('ax' is connected to 'a'). Another rectangular waveform V_N with V_1 as peak value is generated at the output of multiplexer M_1. The ON time δ_T of this rectangular waveform V_N is given as

$$\delta_T = \frac{V_O}{V_R}T \tag{3.35}$$

The R_3C_3 low pass filter gives the average value of this pulse train V_N and is given as

$$V_X = \frac{1}{T}\int_0^{\delta_T} V_1 dt = \frac{V_1}{T}\delta_T$$

$$V_X = \frac{V_1 V_O}{V_R} \tag{3.36}$$

where V_R is a constant value.

The op-amp OA_3 is kept in a negative closed loop configuration, and a positive dc voltage is ensured in the feedback. Hence its inverting terminal voltage will be equal to its non-inverting terminal voltage, i.e.,

$$V_X = V_2 \tag{3.37}$$

From equations (3.36) and (3.37),

$$V_O = \frac{V_2}{V_1} V_R \tag{3.38}$$

Figure 3.15(b) shows re-trigger monosatable as a divider, and its associated waveforms are shown in Figure 3.16(b).

Chapter 4

Time Division Dividers (TDD)—Switching

4.1 SAW TOOTH WAVE BASED TIME DIVISION DIVIDERS

The circuit diagrams of saw tooth wave based time division dividers are shown in Figure 4.1, and their associated waveforms are shown in Figure 4.2. A saw tooth wave V_{S1} of peak value V_R and time period T is generated by

In the circuits of Figure 4.1, the comparator OA_2 compares the saw tooth wave V_{S1} of peak value V_R with the input voltage V_1 and produces a rectangular waveform V_M at its output. The ON time δ_T of this rectangular waveform V_M is given as

$$\delta_T = \frac{V_1}{V_R}T \tag{4.1}$$

The rectangular pulse V_M controls the switch S_1. When V_M is HIGH, the output voltage V_O is connected to the R_3C_2 low pass filter (switch S_1 is closed). When V_M is LOW, zero voltage exists on the R_3C_2 low pass filter (switch S_1 is opened). Another rectangular pulse V_N with maximum value of V_O is generated at the switch S_1 output. The R_3C_2 low pass filter gives the average value of this pulse train V_N and is given as

$$V_X = \frac{1}{T}\int_0^{\delta_T} V_O dt \tag{4.2}$$

$$V_X = \frac{V_O}{T}\delta_T \tag{4.3}$$

DOI: 10.1201/9781003362968-4

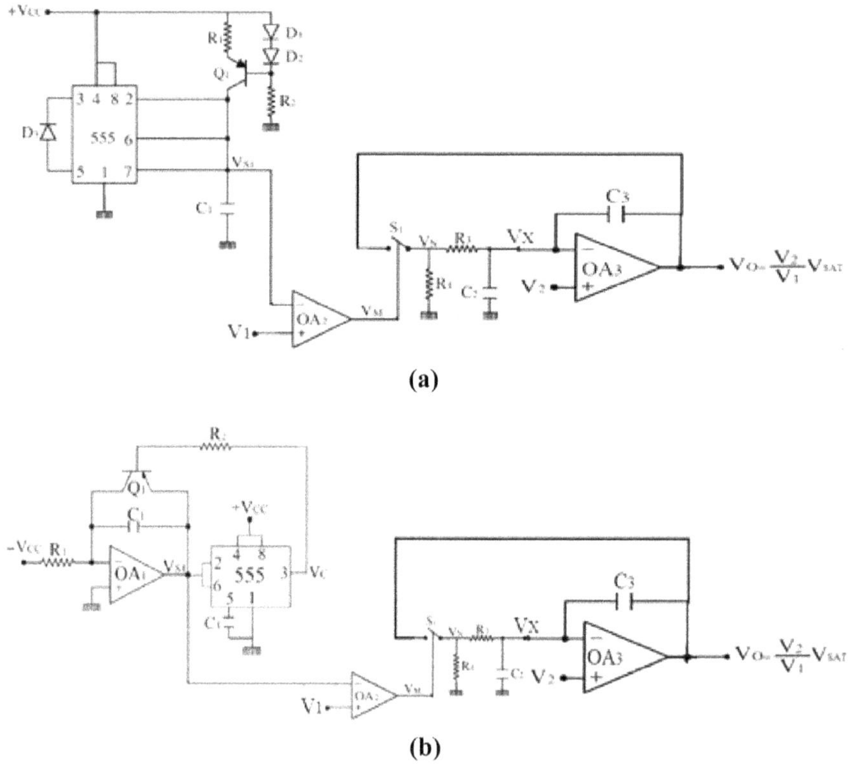

Figure 4.1 (a) Saw tooth wave based time division divider—type I. (b) Saw tooth wave based time division divider—type II.

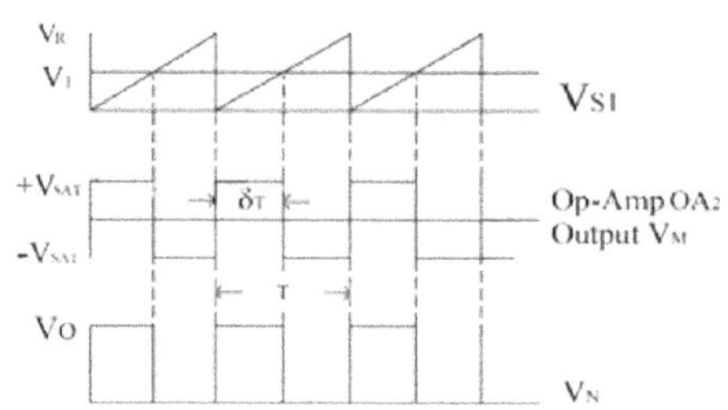

Figure 4.2 Associated waveforms of Figure 4.1.

Equation (4.1) in (4.3) gives

$$V_X = \frac{V_1 V_O}{V_R} \tag{4.4}$$

where $V_R = 2/3\ V_{CC}$.

The op-amp OA_3 is kept in a negative closed loop configuration, and a positive dc voltage is ensured in the feedback. Hence its inverting terminal voltage will be equal to its non-inverting terminal voltage, i.e.,

$$V_X = V_2 \tag{4.5}$$

From equations (4.4) and (4.5),

$$V_O = \frac{V_2}{V_1} V_R \tag{4.6}$$

4.2 TRIANGULAR WAVE REFERENCED TIME DIVISION DIVIDERS

The circuit diagrams of triangular wave based dividers are shown in Figure 4.3, and their associated waveforms are shown in Figure 4.4. A triangular wave V_{T1} with a $\pm V_T$ peak to peak value and time period T is generated by the 555 timer. One input voltage V_1 is compared with the generated triangular wave V_{T1} by the comparator on OA_2. An asymmetrical rectangular waveform V_M is generated at the comparator OA_2 output. From the waveforms shown in Figure 4.5, it is observed that

$$T_1 = \frac{V_T - V_1}{2V_T} T, T_2 = \frac{V_T + V_1}{2V_T} T, T = T_1 + T_2 \tag{4.7}$$

This rectangular wave V_M is given as control input to the switch S_1. During T_2 of V_M, the switch S_1 is closed, and the op-amp OA_3 will work as a non-inverting amplifier. $+V_O$ will be its output, i.e., $V_N = +V_O$. During T_1 of V_M, the switch S_1 is opened, and the op-amp will work as an inverting amplifier. $-V_O$ will be at its output, i.e., $V_N = -V_O$. Another rectangular asymmetrical wave V_N with a peak to peak value of $\pm V_O$ is generated at the op-amp OA_3 output. The R_4C_3 low pass filter gives an average value of the pulse train V_N and is given as

$$V_X = \frac{1}{T}\left[\int_O^{T_2} V_O \, dt + \int_{T_2}^{T_1+T_2} (-V_O)\, dt\right] = \frac{V_O}{T}(T_2 - T_1) \tag{4.8}$$

Equation (4.7) in (4.8) gives

$$V_X = \frac{V_1 V_O}{V_T} \tag{4.9}$$

where $V_T = V_{CC}/3$. (4.10)

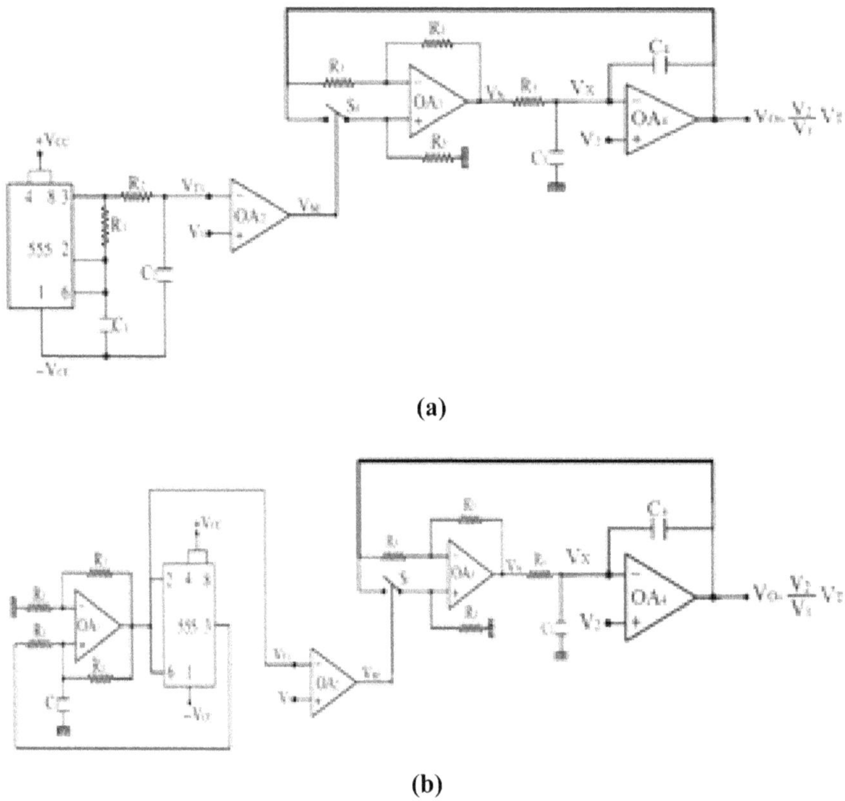

Figure 4.3 (a) Triangular wave based divider—type I. (b) Triangular wave based divider—type II.

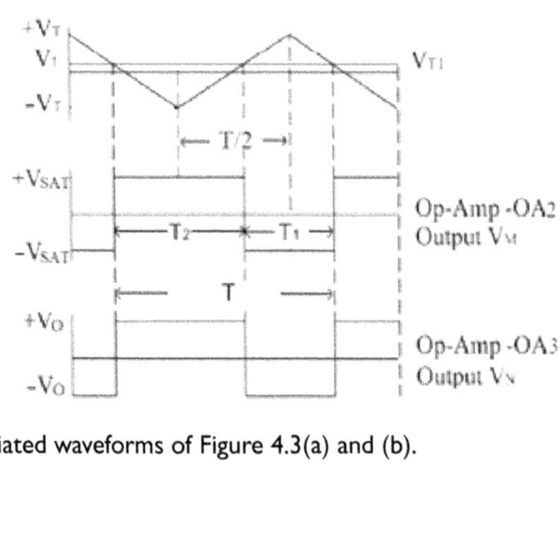

Figure 4.4 Associated waveforms of Figure 4.3(a) and (b).

The op-amp OA_4 is kept in a negative closed loop configuration, and a positive dc voltage is ensured in the feedback. Hence its inverting terminal voltage will be equal to its non-inverting terminal voltage, i.e.,

$$V_X = V_2 \tag{4.11}$$

From equations (4.9) and (4.11),

$$V_O = \frac{V_2}{V_1} V_T \tag{4.12}$$

4.3 TIME DIVISION DIVIDER WITH NO REFERENCE—TYPE I

The divider using the time division principle without using any reference clock is shown in Figure 4.5, and its associated waveforms are shown in Figure 4.6. Initially the 555 timer output is HIGH. The switch S_1 is closed, and the op-amp OA_3 will work as a non-inverting amplifier. $-V_1$ is given to the differential integrator composed of resistor R_1, capacitor C_1, and op-amp OA_1. The output of differential integrator will be

$$V_{T1} = \frac{1}{R_1C_1}\int (V_2 + V_1)dt$$

$$V_{T1} = \frac{(V_1 + V_2)}{R_1C_1} t \tag{4.13}$$

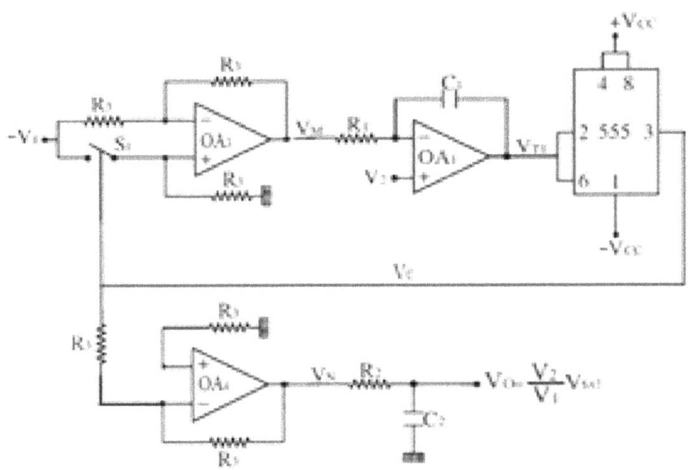

Figure 4.5 Time division divider without reference clock.

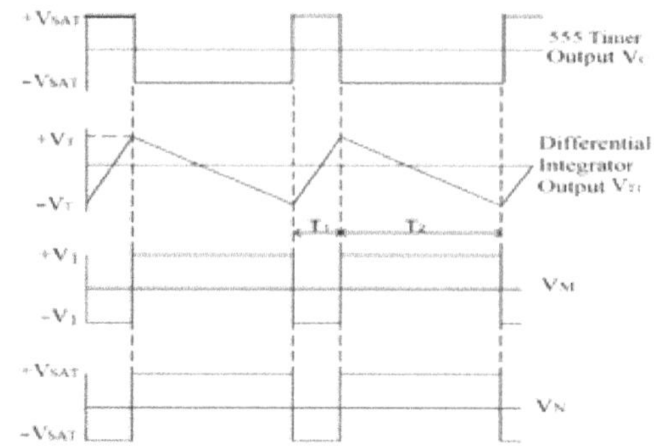

Figure 4.6 Associated waveforms of Figure 4.5.

The output of the differential integrator rises toward positive saturation, and when it reaches the voltage level of $+V_T$, the 555 timer output becomes LOW. The switch S_1 is opened, and the op-amp OA_3 will work as an inverting amplifier. $+V_1$ is given to the differential integrator composed of resistor R_1, capacitor C_1, and op-amp OA_1. Now the output of differential integrator will be

$$V_{T1} = \frac{1}{R_1C_1}\int (V_2 - V_1)dt$$

$$V_{T1} = -\frac{(V_1 - V_2)}{R_1C_1}t \tag{4.14}$$

The output of the differential integrator reverses toward negative saturation, and when it reaches the voltage level $-V_T$, the 555 timer output becomes HIGH, and the cycle therefore repeats, to give an asymmetrical rectangular wave V_C at the output of the 555 timer.

$$V_T = \frac{V_{CC}}{3} \tag{4.15}$$

From the waveforms shown in Figure 4.6, it is observed that

$$T_1 = \frac{V_1 - V_2}{2V_1}T, T_2 = \frac{V_1 + V_2}{2V_1}T, T = T_1 + T_2 \tag{4.16}$$

The 555 timer output is given to the inverting amplifier OA_4. Another rectangular wave V_N is generated at the op-amp OA_4 output with a $\pm V_{SAT}$ peak to peak value. The R_2C_2 low pass filter gives the average value of this pulse train V_N and is given as

$$V_O = \frac{1}{T}\left[\int_{O}^{T_2} V_{SAT}\,dt + \int_{T_2}^{T_1+T_2} (-V_{SAT})\,dt\right]$$

$$V_O = \frac{V_{SAT}(T_2 - T_1)}{T} \tag{4.17}$$

Equation (4.16) in (4.17) gives

$$V_O = \frac{V_2}{V_1} V_{SAT} \tag{4.18}$$

4.4 TIME DIVISION DIVIDER WITH NO REFERENCE—TYPE II

The MCDs using the time division principle without using any reference clock is shown in Figure 4.7, and its associated waveforms are shown in Figure 4.8.

Initially the 555 timer output is HIGH. The inverting amplifier OA_3 output will be LOW. The output of the differential integrator will be

$$V_{T1} = \frac{1}{R_1C_1}\int (V_O + V_{SAT})dt$$

$$V_{T1} = \frac{(V_O + V_{SAT})}{R_1C_1}t \tag{4.19}$$

The output of the differential integrator rises toward positive saturation, and when it reaches the voltage level of $+V_T$, the 555 timer output becomes LOW. The inverting amplifier OA_3 output will be HIGH. The output of differential integrator will now be

$$V_{T1} = \frac{1}{R_1C_1}\int (V_O - V_{SAT})dt$$

$$V_{T1} = -\frac{(V_{SAT} - V_O)}{R_1C_1}t \tag{4.20}$$

The output of differential integrator reverses toward negative saturation, and when it reaches the voltage level $-V_T$, the 555 timer output becomes HIGH, and the cycle therefore repeats, to give an asymmetrical rectangular wave V_C at the output of the 555 timer.

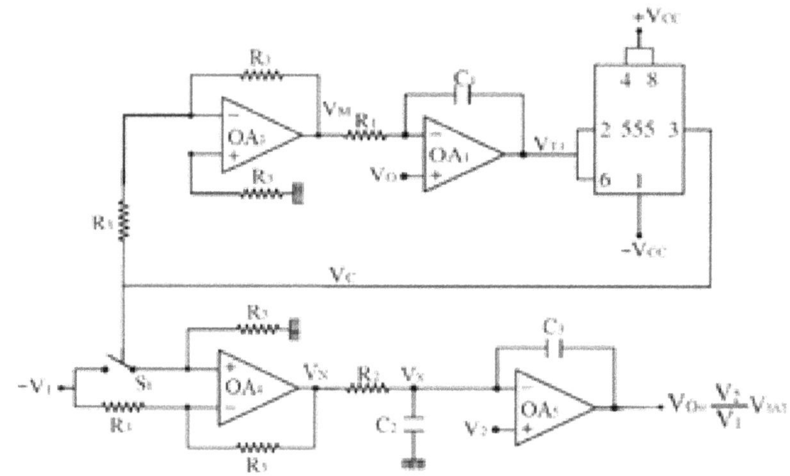

Figure 4.7 Divider without reference clock—II.

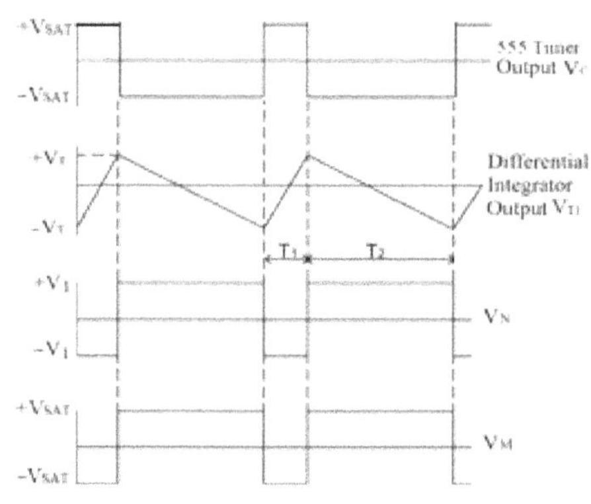

Figure 4.8 Associated waveforms of Figure 4.7.

$$V_T = \frac{V_{CC}}{3} \tag{4.21}$$

From the waveforms shown in Figure 4.8, it is observed that

$$T_1 = \frac{V_{SAT} - V_O}{2V_{SAT}} T, T_2 = \frac{V_{SAT} + V_O}{2V_{SAT}} T, T = T_1 + T_2 \tag{4.22}$$

During the HIGH of V_C, the S_1 is closed, the op-amp OA_4 will work as a non-inverting amplifier, and $-V_1$ is given to the low pass filter. During the LOW of V_C, the S_1 is opened, the op-amp OA_2 will work as an inverting amplifier, and $+V_1$ is given to the low pass filter. Another rectangular wave V_N with $\pm V_1$ as the peak to peak value is generated at the output of op-amp OA_4. The R_2C_2 low pass filter gives the average value of this pulse train V_N and is given as

$$V_X = \frac{1}{T}\left[\int_O^{T_2} V_1\,dt + \int_{T_2}^{T_1+T_2} (-V_1)\,dt\right]$$

$$V_X = \frac{V_1(T_2 - T_1)}{T} \tag{4.23}$$

Equations (4.22) in (4.23) gives

$$V_X = \frac{V_O V_1}{V_{SAT}} \tag{4.24}$$

The op-amp OA_5 is in a negative closed loop configuration, and a positive dc voltage is ensured in the feedback loop. Hence its non-inverting terminal voltage is equal to its inverting terminal voltage, i.e.,

$$V_2 = V_X \tag{4.25}$$

From equations (4.24) and (4.25),

$$V_O = \frac{V_2}{V_1} V_{SAT} \tag{4.26}$$

4.5 DIVIDER FROM 555 ASTABLE MULTIVIBRATOR

4.5.1 Type I

The circuit diagram of a divider using the 555 astable multivibrator is shown in Figure 4.9, and its associated waveforms are shown in Figure 4.10. Refer to the internal diagram of the 555 timer IC shown in Figure 0.1. Initially when we switch on the power supply, the output of the upper comparator CMP_1 will be LOW, i.e., R = 0, and the output of the lower comparator CMP_2 will be HIGH, i.e., S = 1. The flip flop outputs are Q = 1 and Q′ = 0. The timer output at pin 3 will be HIGH, transistor Q_1 is OFF, and hence the discharge pin 7 is at the open position.

The capacitor C_1 is charging toward V_1 through the resistors R_1 and R_2 with a time constant of $(R_1+R_2)C_1$, and its voltage is rising exponentially. When the capacitor voltage is rising above the voltage 2/3 V_{CC}, the output of the upper comparator CMP_1 becomes HIGH, i.e., R = 1, and the output of

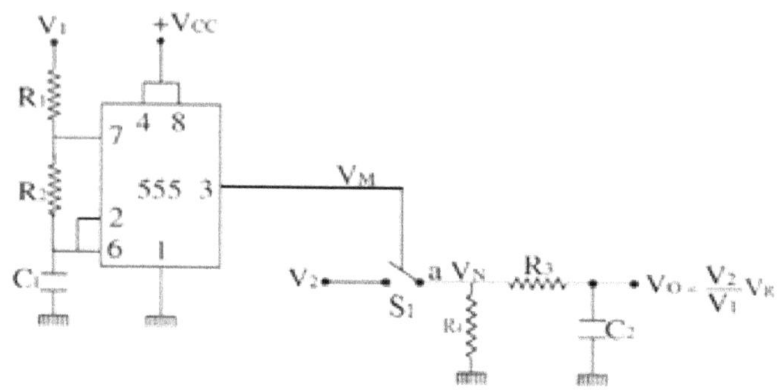

Figure 4.9 Divider from 555 astable.

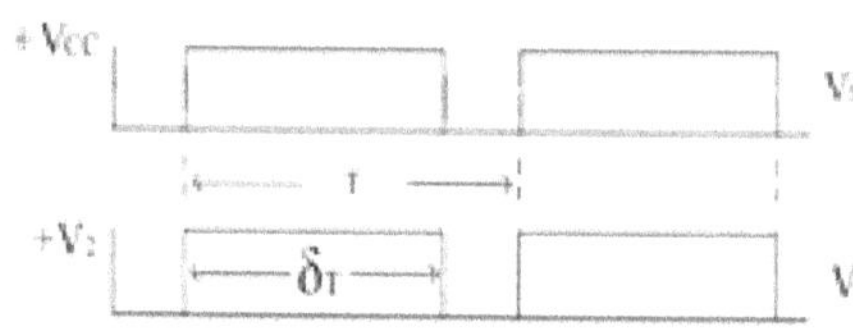

Figure 4.10 Associated waveforms of Figure 4.9.

the lower comparator CMP_2 becomes LOW, i.e., S = 0. The flip flop outputs are Q = 0 and Q′ = 1. The timer output at pin 3 will be LOW, transistor Q_1 is ON, and hence the discharge pin 7 is at GND potential. Now the capacitor C_1 is discharging to GND potential through the resistor R_2 with a time constant of R_2C. When the capacitor voltage falls below 1/3 V_{CC}, the output of the upper comparator CMP_1 becomes LOW, i.e., R = 0, and the output of the lower comparator CMP_2 becomes HIGH, i.e., S = 1. The flip flop outputs are Q = 1 and Q′ = 0. The timer output at pin 3 will be HIGH, transistor Q_1 is OFF, and hence the discharge pin 7 is at the open position.

Now the capacitor starts charging toward V_1, and the cycle therefore repeats to produce periodic pulses at the output pin 3 of the 555 timer.

The ON time of the 555 timer output V_M is inversely proportional to V_1. During the ON time δ_T, the second input voltage V_2 is connected to the R_3C_2 low pass filter (switch S_1 is closed). During the OFF time of V_M, zero voltage is connected to the R_3C_2 low pass filter (switch S_1 is opened). Another rectangular waveform V_N with V_2 as the peak value is generated at the output of switch S_1. The ON time δ_T of this rectangular pulse V_N is given as

$$\delta_T = \frac{V_R}{V_1} T \tag{4.27}$$

where V_R is a constant value.

The R_3C_2 low pass filter gives the average value of this pulse train V_N and is given as

$$V_O = \frac{1}{T}\int_0^{\delta_T} V_2 dt = \frac{V_2}{T}\delta_T$$

$$V_O = \frac{V_2}{V_1}V_R \qquad (4.28)$$

4.5.2 Divider from 555 Astable Multivibrator—Type II

The circuit diagram of a divider using the 555 timer astable multivibrator is shown in Figure 4.11, and its associated waveforms are shown in Figure 4.12. Refer to the internal diagram of the 555 timer IC shown in Figure 0.1. Initially when we switch on the power supply, the output of the upper comparator CMP_1 will be LOW, i.e., R = 0, and the output of the lower comparator CMP_2 will be HIGH, i.e., S = 1. The flip flop outputs are Q = 1 and Q′ = 0. The timer output at pin 3 will be HIGH, transistor Q_1 is OFF, and hence the discharge pin 7 is at the open position.

The capacitor C_1 is charging toward $+V_{CC}$ through the resistors R_1 and R_2 with a time constant of $(R_1+R_2)C_1$, and its voltage is rising

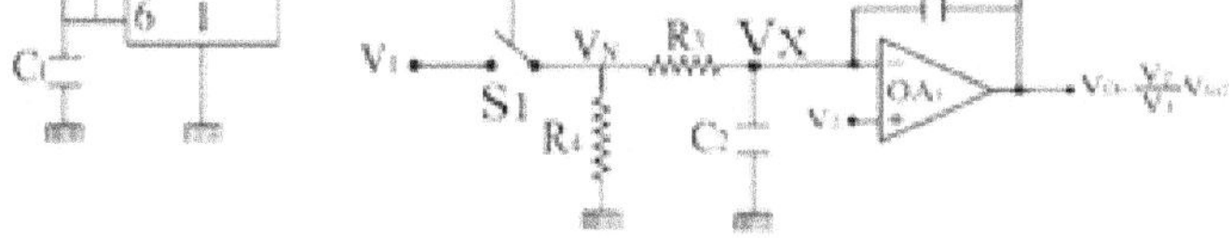

Figure 4.11 Divider with 555 timer astable multivibrator.

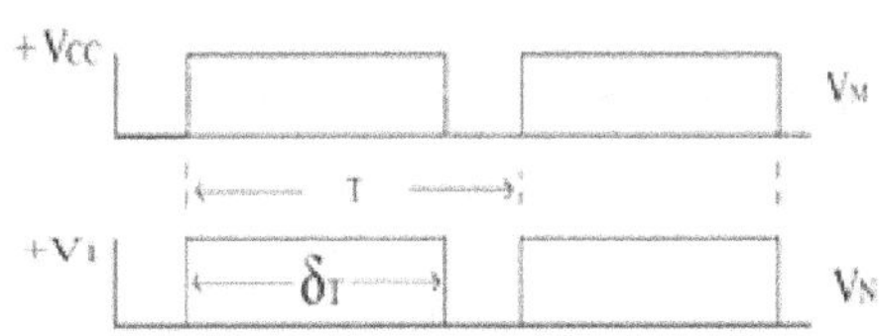

Figure 4.12 Associated waveforms of Figure 4.11.

exponentially. When the capacitor voltage is rising above the voltage V_O, the output of the upper comparator CMP_1 becomes HIGH, i.e., R = 1, and the output of the lower comparator CMP_2 becomes LOW, i.e., S = 0. The flip flop outputs are Q = 0 and Q′ = 1. The timer output at pin 3 will be LOW, transistor Q_1 is ON, and hence the discharge pin 7 is at GND potential. Now the capacitor C_1 is discharging to GND potential through the resistor R_2 with a time constant of R_2C. When the capacitor voltage falls below 1/3 V_{CC}, the output of the upper comparator CMP_1 becomes LOW, i.e., R = 0, the output of the lower comparator CMP_2 becomes HIGH, i.e., S = 1. The flip flop outputs are Q = 1 and Q′ = 0. The timer output at pin 3 will be HIGH, transistor Q_1 is OFF, and hence the discharge pin 7 is at the open position.

Now the capacitor starts charging toward $+V_{CC}$, and the cycle therefore repeats to produce periodic pulses at the output pin 3 of the 555 timer.

The ON time of the 555 timer output V_M is proportional to V_O, which is applied at its pin 5. During the ON time δ_T, the input voltage V_1 is connected to the R_3C_2 low pass filter (switch S_1 is closed). During the OFF time of V_M, zero voltage exists on the R_3C_2 low pass filter (switch S_1 is opened). Another rectangular waveform V_N with V_1 as the peak value is generated at the emitter of transistor. The ON time of V_M is given as

$$\delta_T = \frac{V_O}{V_R} T \tag{4.29}$$

The R_3C_2 low pass filter gives the average value of this pulse train V_N and is given as

$$V_X = \frac{1}{T}\int_0^{\delta_T} V_1 dt = \frac{V_1}{T}\delta_T$$

$$V_X = \frac{V_1 V_O}{V_R} \tag{4.30}$$

where V_R is a constant value.

The op-amp OA_1 is kept in a negative closed loop configuration, and a positive dc voltage is ensured in the feedback. Hence its inverting terminal voltage will be equal to its non-inverting terminal voltage, i.e.,

$$V_X = V_2 \tag{4.31}$$

From equations (4.30) and (4.31),

$$V_O = \frac{V_2}{V_1} V_R \tag{4.32}$$

4.6 DIVIDER FROM 555 MONOSTABLE MULTIVIBRATOR

4.6.1 Type I

The circuit diagram of a divider using the 555 monostable multivibrator is shown in Figure 4.13, and its associated waveforms are shown in Figure 4.14. Refer to the internal diagram of the 555 timer IC shown in Figure 0.1. Initially when the power supply is switched on, the output of the upper comparator CMP_1 will be LOW, i.e., R = 0, and the output of the lower comparator CMP_2 will be HIGH, i.e., S = 1. The flip flop outputs are Q = 1 and Q′ = 0. The timer output at pin 3 will be HIGH, transistor Q_1 is OFF, and hence the discharge pin 7 is at the open position. The capacitor C_1 is charging toward V_1 through the resistor R_1. The capacitor voltage is rising exponentially, and when it reaches the value of 2/3 V_{CC}, the output of the upper comparator CMP_1 becomes HIGH, i.e., R = 1, and the output of the lower comparator CMP_2 becomes LOW, i.e., S = 0. The flip flop outputs are Q = 0 and Q′ = 1. The timer output at pin 3 will be LOW, transistor Q_1 is ON, and hence the discharge pin 7 is at GND potential. Now the capacitor C_1 is short circuited, zero voltage exists at pin 6, and the output of the upper comparator CMP_1 becomes LOW, i.e., R = 0. A trigger pulse is applied at pin 2, and when the trigger voltage comes down to 1/3 V_{CC}, the output of the lower comparator CMP_2 becomes HIGH, i.e., S = 1. The flip flop outputs are Q = 1 and Q′ = 0. The timer output at pin 3 will be HIGH, transistor Q_1 is OFF, and hence the discharge pin 7 is at the open position.

Now the capacitor C_1 is charging toward V_1, and the sequence therefore repeats for every trigger input pulse.

The ON time of the 555 timer output V_M is inversely proportional to V_1. The output of the 555 timer controls switch S_1. During the ON time δ_T, the second input voltage V_2 is connected to the R_3C_3 low pass filter (switch S_1 is closed). During the OFF time of V_M, zero voltage exist on the R_3C_3 low pass filter (switch S_1 is opened). Another rectangular waveform V_N with V_2 as the peak value is generated at the output of switch S_1. The ON time of rectangular wave is given as

$$\delta_T = \frac{V_R}{V_1} T \tag{4.33}$$

The R_3C_3 low pass filter gives the average value of this pulse train V_N and is given as

$$V_O = \frac{1}{T}\int_0^{\delta_T} V_2 dt = \frac{V_2}{T}\delta_T$$

$$V_O = \frac{V_2}{V_1} V_R \tag{4.34}$$

where V_R is a constant value.

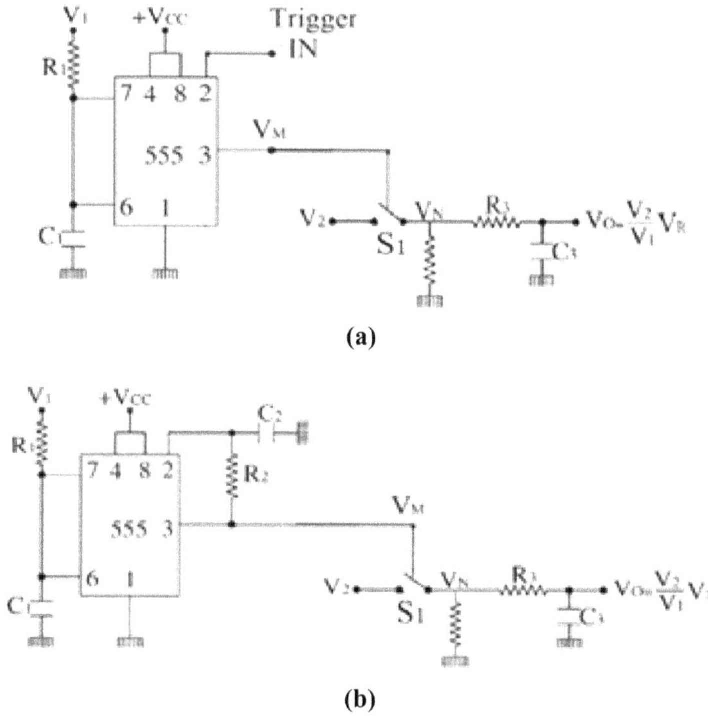

Figure 4.13 (a) Divider using 555 timer monostable multivibrator. (b) Divider using auto-trigger monostable multivibrator.

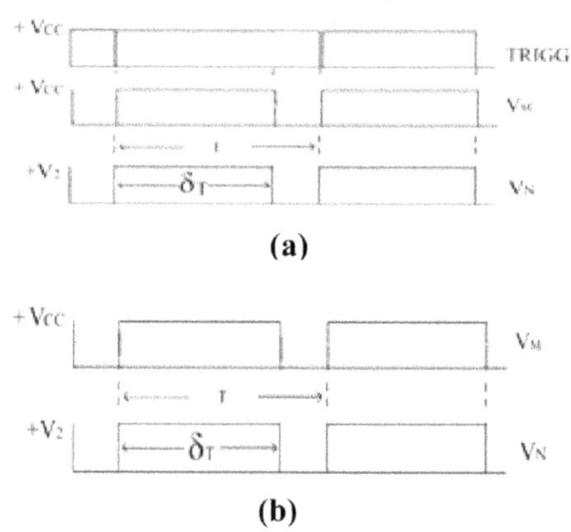

Figure 4.14 (a) Associated waveforms of Figure 4.13(a). (b) Associated waveforms of Figure 4.13(b).

Figure 4.19(b) shows the auto-trigger monostable multivibrator used as an analog divider.

4.6.2 Divider from 555 Monostable Multivibrator

The circuit diagram of divider using the 555 timer monostable multivibrator is shown in Figure 4.15, and its associated waveforms are shown in Figure 4.16. Refer to the internal diagram of the 555 timer IC shown in Figure 0.1. Initially when the power supply is switched on, the output of the upper comparator CMP_1 will be LOW, i.e., R = 0, and the output of the lower comparator CMP_2 will be HIGH, i.e., S = 1. The flip flop outputs are Q = 1 and Q′ = 0. The timer output at pin 3 will be HIGH, transistor Q_1 is OFF, and hence the discharge pin 7 is at the open position. The capacitor C_1 is charging toward $+V_{CC}$ through the resistor R_1. The capacitor voltage is rising exponentially, and when it reaches the value of V_O, the output of the upper comparator CMP_1 becomes HIGH, i.e., R = 1, and the output of the lower comparator CMP_2 becomes LOW, i.e., S = 0. The flip flop outputs are Q = 0 and Q′ = 1. The timer output at pin 3 will be LOW, transistor Q_1 is ON, and hence the discharge pin 7 is at GND potential. Now the capacitor C_1 is short circuited, zero voltage exists at pin 6, and the output of the upper comparator CMP_1 becomes LOW, i.e., R = 0. A trigger pulse is applied at pin 2, and when the trigger voltage comes down to 1/3 V_{CC}, the output of the

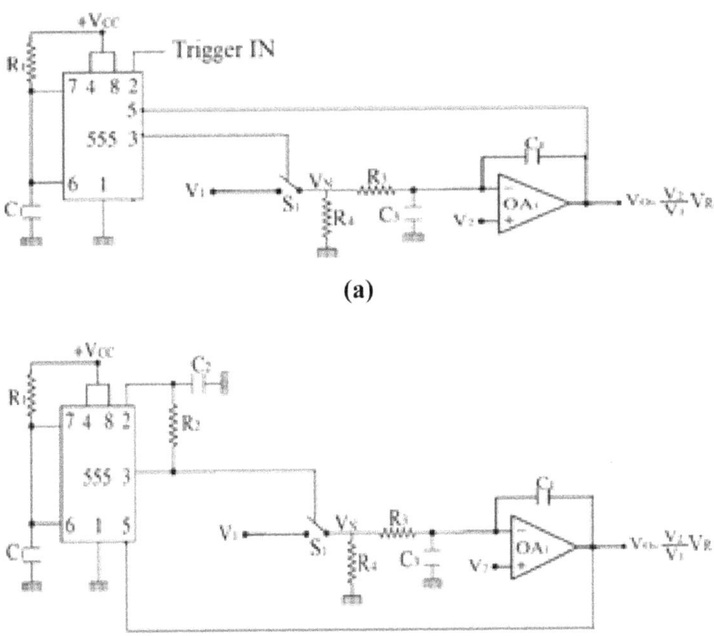

Figure 4.15 (a) Divider from 555 monostable. (b) Divider with 555 re-trigger monostable multivibrator.

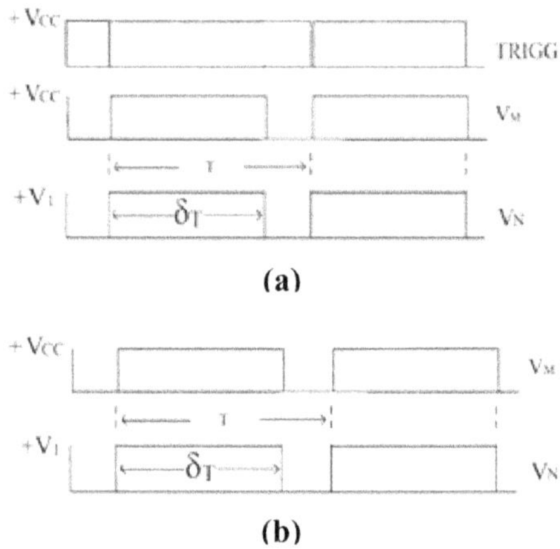

Figure 4.16 (a) Associated waveforms of Figure 4.15(a). (b) Associated waveforms of Figure 4.15(b).

lower comparator CMP_2 becomes HIGH, i.e., S = 1. The flip flop outputs are Q = 1 and Q′ = 0. The timer output at pin 3 will be HIGH, transistor Q_1 is OFF, and hence the discharge pin 7 is at the open position.

Now the capacitor C_1 is charging toward $+V_{CC}$, and the sequence therefore repeats for every trigger input pulse.

The ON time of the 555 timer output V_M is proportional to V_O which is applied at its pin 5. The 555 timer output controls switch S1. During the ON time δ_T, the first input voltage V_1 is connected to the R_3C_3 low pass filter (switch S1 is closed). During the OFF time of V_M, zero voltage exists on the R_3C_2 low pass filter (switch S_1 is opened). Another rectangular waveform V_N with V_1 as the peak value is generated at the output of switch S_1. The ON time δ_T of this rectangular waveform V_N is given as

$$\delta_T = \frac{V_O}{V_R} T \tag{4.35}$$

The R_3C_3 low pass filter gives the average value of this pulse train V_N and is given as

$$V_X = \frac{1}{T}\int_0^{\delta_T} V_1 dt = \frac{V_1}{T}\delta_T$$

$$V_X = \frac{V_1 V_O}{V_R} \tag{4.36}$$

where V_R is a constant value.

The op-amp OA_1 is kept in a negative closed loop configuration, and a positive dc voltage is ensured in the feedback. Hence its inverting terminal voltage will be equal to its non-inverting terminal voltage, i.e.,

$$V_X = V_2 \tag{4.37}$$

From equations (4.36) and (4.37),

$$V_O = \frac{V_2}{V_1} V_R \tag{4.38}$$

Figure 4.15(b) shows a divider using the 555 re-trigger monostable multivibrator.

Chapter 5

Time Division Multipliers cum Dividers (MCDs)—Multiplexing

5.1 SAW TOOTH WAVE REFERENCED MCDS

5.1.1 Saw Tooth Wave Referenced MCD—Type I—Double Multiplexing and Averaging

The circuit diagram of double multiplexing–averaging time division MCD is shown in Figure 5.1, and its associated waveforms are shown in Figure 5.2. A saw tooth wave V_{S1} of peak value V_R and time period T is generated by the 555 timer.

The comparator OA_3 compares the saw tooth wave V_{S1} with the voltage V_Y and produces a rectangular waveform V_K. The ON time δ_T of V_K, is given as

$$\delta_T = \frac{V_Y}{V_R} T \tag{5.1}$$

The rectangular pulse V_K controls the second multiplexer M_2. When V_K is HIGH, the first input voltage V_1 is connected to the R_4C_3 low pass filter ('by' is connected to 'b'). When V_K is LOW, zero voltage is connected to the R_4C_3 low pass filter ('bx' is connected to 'b'). Another rectangular pulse V_M with maximum value of V_1 is generated at the multiplexer M_2 output. The R_4C_3 low pass filter gives the average value of this pulse train V_M and is given as

$$V_X = \frac{1}{T}\int_0^{\delta_T} V_1 dt = \frac{V_1}{T}\delta_T \tag{5.2}$$

$$V_X = \frac{V_1 V_Y}{V_R} \tag{5.3}$$

The op-amp OA_4 is configured in a negative closed loop feedback, and a positive dc voltage is ensured in the feedback loop. Hence its inverting terminal voltage must be equal to its non-inverting terminal voltage.

DOI: 10.1201/9781003362968-5

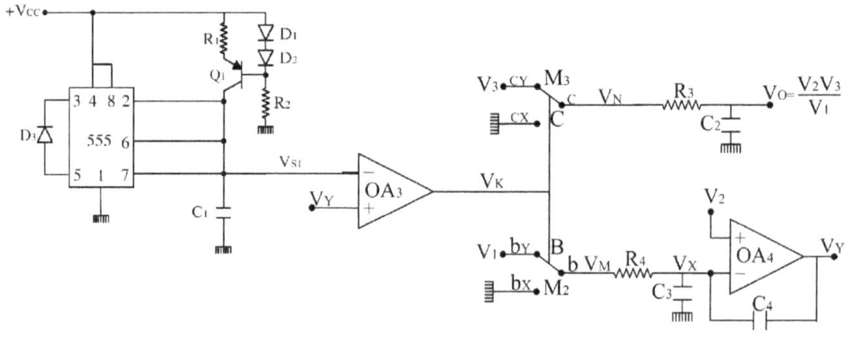

Figure 5.1 Double multiplexing—averaging MCD.

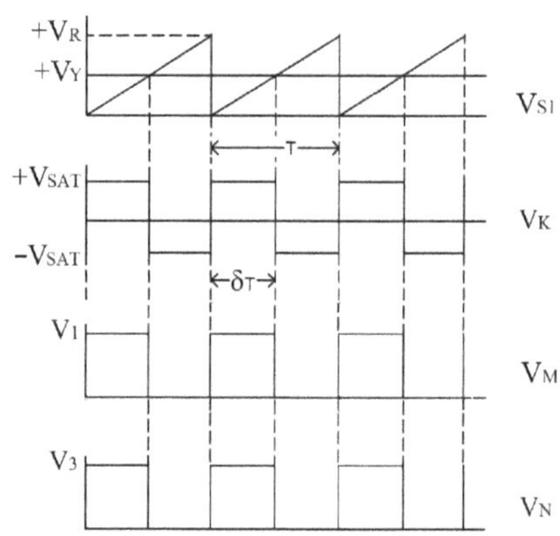

Figure 5.2 Associated waveforms of Figure 5.1.

$$V_2 = V_X \tag{5.4}$$

From equations (5.3) and (5.4),

$$V_Y = \frac{V_2 V_R}{V_1} \tag{5.5}$$

The rectangular pulse V_K also controls the third multiplexer M_3. When V_K is HIGH, the third input voltage V_3 is connected to the R_3C_2 low pass filter ('cy' is connected to 'c'). When V_K is LOW, zero voltage is connected to the R_3C_2 low pass filter ('cx' is connected to 'c'). Another rectangular pulse V_N with a maximum value of V_3 is generated at the multiplexer M_3 output. The R_3C_2 low pass filter gives the average value of this pulse train V_N and is given as

$$V_O = \frac{1}{T}\int_0^{\delta_T} V_3 dt = \frac{V_3}{T}\delta_T \tag{5.6}$$

Equations (5.1) and (5.5) in (5.6) gives

$$V_O = \frac{V_2 V_3}{V_1} \tag{5.7}$$

5.1.2 Saw Tooth Based MCD—Type II

The circuit diagram of the saw tooth wave based time division multiply-divide MCD is shown in Figure 5.3, and its associated waveforms are shown in Figure 5.4. A saw tooth wave V_{S1} of peak value V_R is generated by the 555 timer.

The comparator OA_2 compares the saw tooth wave V_{S1} with an input voltage V_3 and produces a rectangular wave form V_M. The ON time δ_{T1} of V_M, is given as

$$\delta_{T1} = \frac{V_3}{V_R} T \tag{5.8}$$

The rectangular pulse V_M controls the second multiplexer M_2. When V_M is HIGH, the second input voltage V_2 is connected to the R_4C_2 low pass filter ('by' is connected to 'b'). When V_M is LOW, zero voltage is connected to the R_4C_2 low pass filter ('bx' is connected to 'b'). Another rectangular pulse V_K with a maximum value of V_2 is generated at the multiplexer M_2 output.

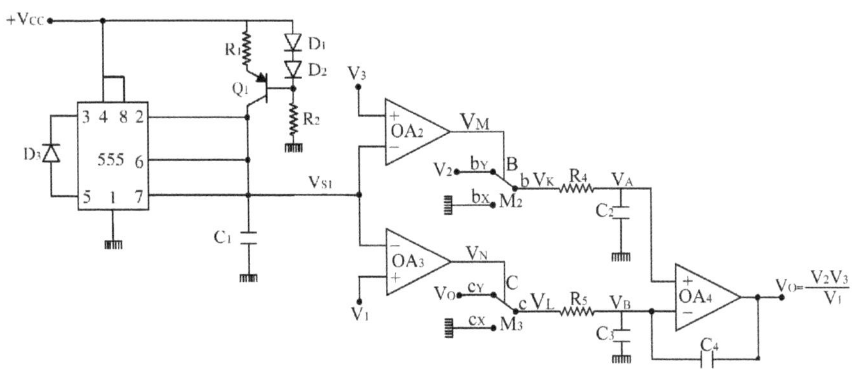

Figure 5.3 Saw tooth wave based time division multiplier—type II.

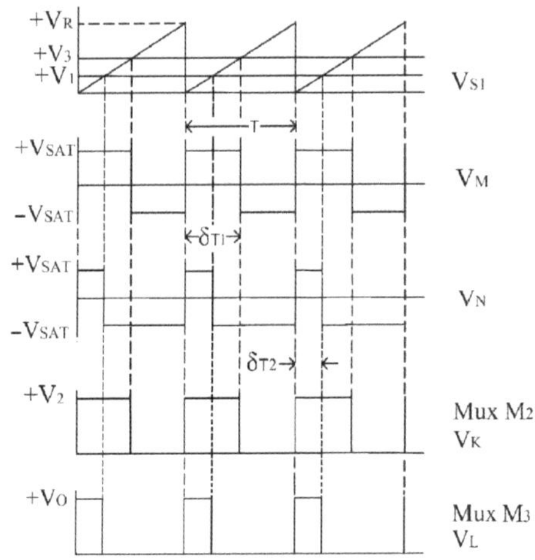

Figure 5.4 Associated wave forms of Figure 5.3.

The R_4C_2 low pass filter gives the average value of this pulse train V_K and is given as

$$V_A = \frac{1}{T}\int_0^{\delta_{T1}} V_2 dt = \frac{V_2}{T}\delta_{T1} \tag{5.9}$$

$$V_A = \frac{V_2 V_3}{V_R} \tag{5.10}$$

The comparator OA_3 compares the saw tooth wave V_{S1} with the first input voltage V_1 and produces a rectangular waveform V_N. The ON time δ_{T2} of V_N, is given as

$$\delta_{T2} = \frac{V_1}{V_R} T \tag{5.11}$$

The rectangular pulse V_N controls the third multiplexer M_3. When V_N is HIGH, the output voltage V_O is connected to the R_5C_3 low pass filter ('cy' is connected to 'c'). When V_N is LOW, zero voltage is connected to the R_5C_3 low pass filter ('cx' is connected to 'c'). Another rectangular pulse V_L with maximum value of V_O is generated at the multiplexer M_3 output. The R_5C_3 low pass filter gives the average value of this pulse train V_L and is given as

$$V_B = \frac{1}{T}\int_0^{\delta_{T2}} V_O dt = \frac{V_O}{T}\delta_{T2}$$

$$V_B = \frac{V_1 V_O}{V_R} \tag{5.12}$$

The op-amp OA_4 is configured in a negative closed loop feedback, and a positive dc voltage is ensured in the feedback loop. Hence its inverting terminal voltage must be equal to its non-inverting terminal voltage, i.e.,

$$V_A = V_B \tag{5.13}$$

Equations (5.10) and (5.12) in (5.13) give

$$V_O = \frac{V_2 V_3}{V_1} \tag{5.14}$$

5.1.3 Saw Tooth Wave Referenced MCD—Type III

The circuit diagram of saw tooth wave based time division divide-multiply MCD is shown in Figure 5.5, and its associated waveforms are shown in Figure 5.6. As discussed in chapter 1, a saw tooth wave V_{S1} with peak value V_R and time period T is generated by the 555 timer.

The comparator OA_2 compares the saw tooth wave with an input voltage V_1 and produces a rectangular waveform V_M. The ON time δ_{T1} of V_M, is given as

$$\delta_{T1} = \frac{V_1}{V_R}T \tag{5.15}$$

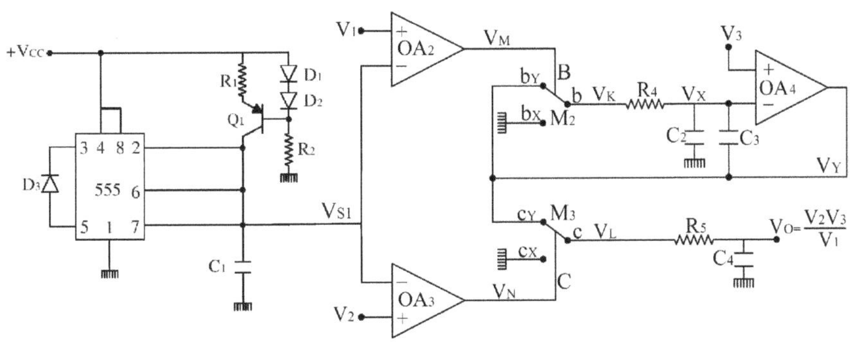

Figure 5.5 Saw tooth wave based time division multiplier—type III.

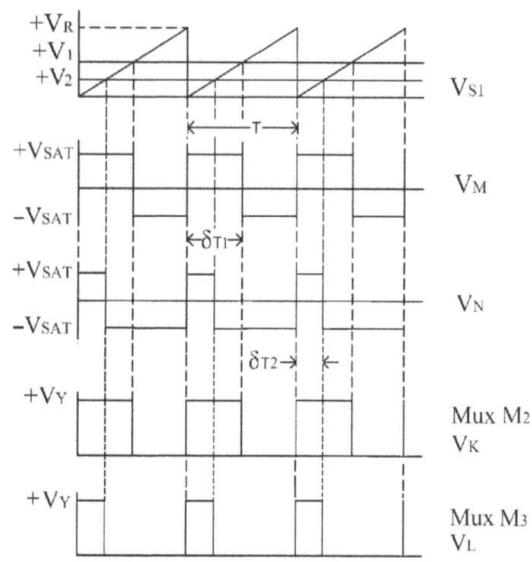

Figure 5.6 Associated waveforms of Figure 5.5.

The rectangular pulse V_M controls the second multiplexer M_2. When V_M is HIGH, the voltage V_Y is connected to the R_4C_2 low pass filter ('by' is connected to 'b'). When V_M is LOW, zero voltage is connected to the R_4C_2 low pass filter ('bx' is connected to 'b'). Another rectangular pulse V_K with a maximum value of V_Y is generated at the multiplexer M_2 output. The R_4C_2 low pass filter gives the average value of this pulse train V_K and is given as

$$V_X = \frac{1}{T}\int_0^{\delta_{T1}} V_Y dt = \frac{V_Y}{T}\delta_{T1}$$

$$V_X = \frac{V_1 V_Y}{V_R} \tag{5.16}$$

The op-amp OA_4 is configured in a negative closed loop feedback, and a positive dc voltage is ensured in the feedback loop. Hence its inverting terminal voltage must be equal to its non-inverting terminal voltage, i.e.,

$$V_X = V_3 \tag{5.17}$$

From equations (5.16) and (5.17),

$$V_Y = \frac{V_3 V_R}{V_1} \tag{5.18}$$

The comparator OA_3 compares the saw tooth wave with the second input voltage V_2 and produces a rectangular waveform V_N. The ON time δ_{T2} of V_N is given as

$$\delta_{T2} = \frac{V_2}{V_R} T \tag{5.19}$$

The rectangular pulse V_N controls the third multiplexer M_3. When V_N is HIGH, the voltage V_Y is connected to the R_5C_4 low pass filter ('cy' is connected to 'c'). When V_N is LOW, zero voltage is connected to the R_5C_4 low pass filter ('cx' is connected to 'c'). Another rectangular pulse V_L with maximum value of V_Y is generated at the multiplexer M_3 output. The R_5C_4 low pass filter gives the average value of this pulse train V_L and is given as

$$V_O = \frac{1}{T} \int_0^{\delta_{T2}} V_Y dt = \frac{V_Y}{T} \delta_{T2} \tag{5.20}$$

$$V_O = \frac{V_2 V_Y}{V_R} \tag{5.21}$$

Equation (5.18) in (5.21) gives

$$V_O = \frac{V_2 V_3}{V_1} \tag{5.22}$$

5.2 TRIANGULAR WAVE BASED TIME DIVISION MCDS

5.2.1 Type I

The circuit diagram of the triangular wave referenced time division multiplier cum divider is shown in Figure 5.7, and its associated waveforms are shown in Figure 5.8. A triangular wave V_{T1} of $\pm V_T$ peak to peak values and time period T is generated by the 555 timer. The comparator OA_3 compares the triangular wave V_{T1} with the voltage V_Y and produces an asymmetrical rectangular wave V_K. From Figure 5.11, it is observed that

$$T_1 = \frac{V_T - V_Y}{2V_T} T, T_2 = \frac{V_T + V_Y}{2V_T} T, T = T_1 + T_2 \tag{5.23}$$

The rectangular wave V_K controls the multiplexer M_1, which connects $+V_1$ to its output during T_2 ('ay' is connected to 'a') and $-V_1$ to its output during T_1 ('ax' is connected to 'a'). Another asymmetrical rectangular waveform V_N

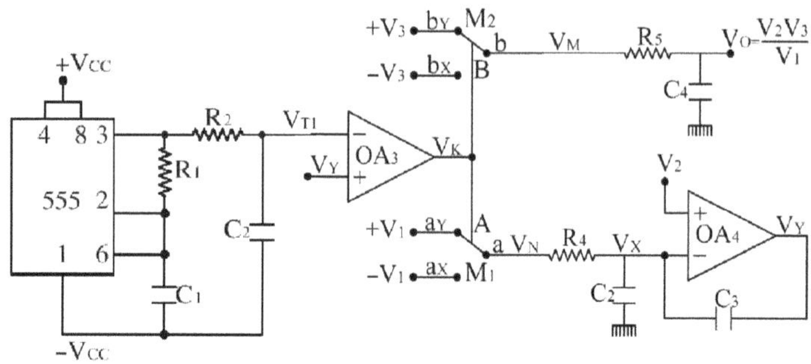

Figure 5.7 Triangular wave based time division MCD—type I.

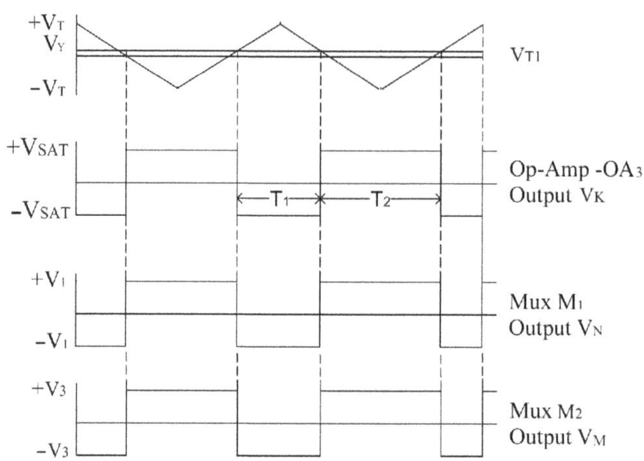

Figure 5.8 Associated waveforms of Figure 5.7.

is generated at the multiplexer M_1 output with $\pm V_1$ peak to peak values. The R_4C_2 low pass filter gives the average value of V_N and is given as

$$V_X = \frac{1}{T}\left[\int_{O}^{T_2} V_1\,dt + \int_{T_2}^{T_1+T_2} (-V_1)\,dt\right] = \frac{V_1}{T}[T_2 - T_1] \tag{5.24}$$

$$V_X = \frac{V_1 V_Y}{V_T}$$

The op-amp OA_4 is configured in a negative closed loop feedback, and a positive dc voltage is ensured in the feedback loop. Hence its inverting terminal voltage must be equal to its non-inverting terminal voltage, i.e.,

$$V_X = V_2 \tag{5.25}$$

From equations (5.24) and (5.25),

$$V_Y = \frac{V_2 V_T}{V_1} \tag{5.26}$$

The rectangular wave V_K also controls the multiplexer M_2, which connects $+V_3$ to its output during T_2 ('by' is connected to 'b') and $-V_3$ to its output during T_1 ('bx' is connected to 'b'). Another asymmetrical rectangular wave V_M is generated at the multiplexer M_2 output with $\pm V_3$ peak to peak values. The R_5C_4 low pass filter gives the average value V_O and is given as

$$V_O = \frac{1}{T}\left[\int_O^{T_2} V_3\,dt + \int_{T_2}^{T_1+T_2} (-V_3)\,dt\right] = \frac{V_3}{T}[T_2 - T_1] \tag{5.27}$$

$$V_O = \frac{V_3 V_Y}{V_T}$$

Equation (5.26) in (5.27) gives

$$V_O = \frac{V_2 V_3}{V_1} \tag{5.28}$$

5.2.2 Triangular Wave Based Time Division MCD—Type II

The circuit diagram of triangular wave based divide-multiply MCD is shown in Figure 5.9, and its associated waveforms are shown in Figure 5.10. A triangular wave V_{T1} of $\pm V_T$ peak to peak values and time period T is generated around the 555 timer IC

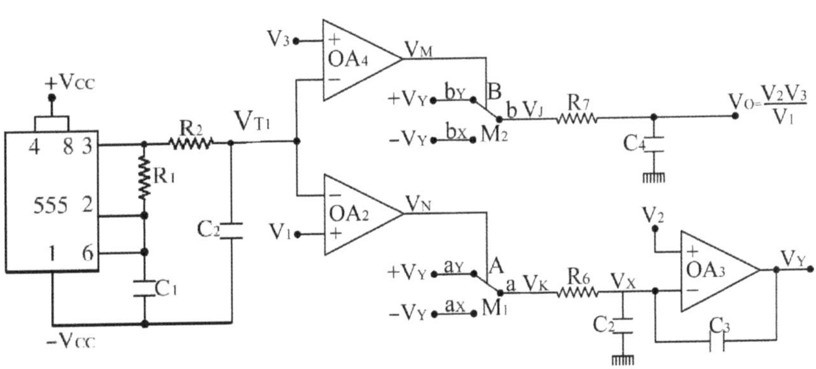

Figure 5.9 Triangular wave based divide-multiply MCD

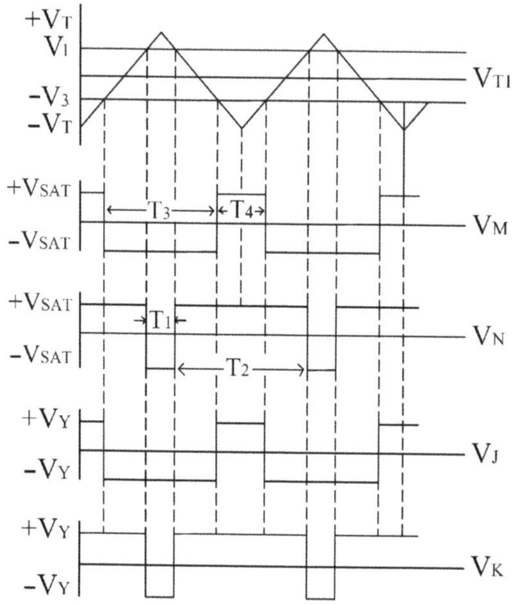

Figure 5.10 Associated waveforms of Figure 5.9

The first input voltage V_1 is compared with the triangular wave V_{T1} by the comparator on OA_2. An asymmetrical rectangular waveform V_N is generated at the comparator OA_2 output. From the waveforms shown in Figure 5.14, it is observed that

$$T_1 = \frac{V_T - V_1}{2V_T} T, T_2 = \frac{V_T + V_1}{2V_T} T, T = T_1 + T_2 \tag{5.29}$$

This rectangular wave V_N is given as control input to the multiplexer M_1. The multiplexer M_1 connects the voltage $+V_Y$ during T_2 ('ay' is connected to 'a') and $-V_Y$ during T_1 ('ax' is connected to 'a'). Another rectangular asymmetrical wave V_K with peak to peak values of $\pm V_Y$ is generated at the multiplexer M_1 output. The R_6C_2 low pass filter gives the average value of the pulse train V_N, which is given as

$$V_X = \frac{1}{T}\left[\int_O^{T_2} V_Y\,dt + \int_{T_2}^{T_1+T_2} (-V_Y)\,dt\right] = \frac{V_Y}{T}(T_2 - T_1)$$

$$V_X = \frac{V_1 V_Y}{V_T} \tag{5.30}$$

The op-amp OA_3 is configured in a negative closed loop feedback, and a positive dc voltage is ensured in the feedback loop. Hence its inverting terminal voltage must be equal to its non-inverting terminal voltage, i.e.,

$$V_2 = V_X \tag{5.31}$$

From equations (5.30) and (5.31),

$$V_Y = \frac{V_2 V_T}{V_1} \tag{5.32}$$

The third input voltage V_3 is compared with the generated triangular wave V_{T1} by the comparator on OA_4. An asymmetrical rectangular waveform V_M is generated at the comparator OA_4 output. From the waveforms shown in Figure 5.10, it is observed that

$$T_3 = \frac{V_T - V_3}{2V_T} T, T_4 = \frac{V_T + V_3}{2V_T} T, T = T_3 + T_4 \tag{5.33}$$

This rectangular wave V_M is given as control input to the multiplexer M_2. The multiplexer M_2 connects the voltage $+V_Y$ during T_4 ('by' is connected to 'b') and $-V_Y$ during T_3 ('bx' is connected to 'b'). Another rectangular asymmetrical wave V_J with peak to peak values of $\pm V_Y$ is generated at the multiplexer M_2 output. The R_7C_4 low pass filter gives the average value of the pulse train V_J and is given as

$$V_O = \frac{1}{T}\left[\int_O^{T_4} V_Y\, dt + \int_{T_4}^{T_3+T_4} (-V_Y) dt\right] = \frac{V_Y}{T}(T_4 - T_3) \tag{5.34}$$

$$V_O = \frac{V_Y V_3}{V_T} \tag{5.35}$$

Equation (5.32) in (5.35) gives

$$V_O = \frac{V_2 V_3}{V_1} \tag{5.36}$$

5.2.3 Triangular Wave Based MCD—Type III

The circuit diagrams of a triangular wave based multiply-divide time division MCD is shown in Figure 5.11, and its associated waveforms are shown in Figure 5.12. A triangular wave of $\pm V_T$ peak to peak value and time period T

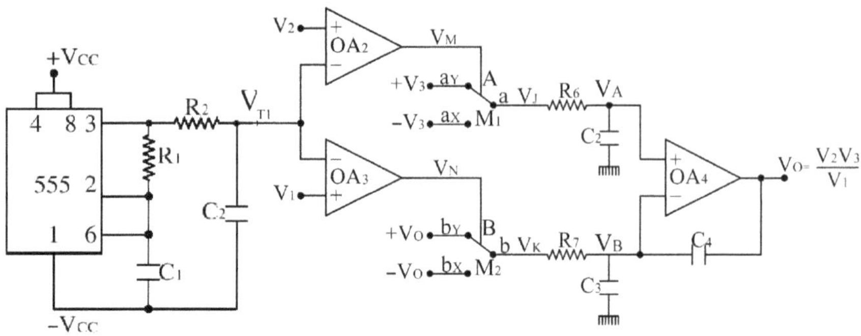

Figure 5.11 Triangular wave based time division multiply-divide MCD—type III.

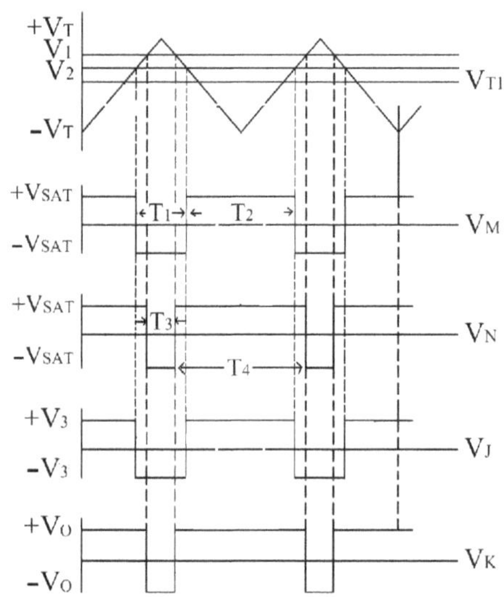

Figure 5.12 Associated waveforms of Figure 5.11.

is generated around the 555 timer. The second input voltage V_2 is compared with the triangular wave V_{T1} by the comparator on OA_2. An asymmetrical rectangular waveform V_M is generated at the comparator OA_2 output. From the waveforms shown in Figure 5.12, it is observed that

$$T_1 = \frac{V_T - V_2}{2V_T}T, \quad T_2 = \frac{V_T + V_2}{2V_T}T, T = T_1 + T_2 \tag{5.37}$$

This rectangular wave V_M is given as control input to the multiplexer M_1. The multiplexer M_1 connects the third input voltage $+V_3$ during T_2 ('ay' is connected to 'a') and $-V_3$ during T_1 ('ax' is connected to 'a'). Another rectangular asymmetrical square wave V_J with peak to peak values of $\pm V_3$ is generated at the multiplexer M_1 output. The R_6C_2 low pass filter gives the average value of the pulse train V_J and is given as

$$V_A = \frac{1}{T}\left[\int_{O}^{T_2} V_3\, dt + \int_{T_2}^{T_1+T_2} (-V_3)\, dt\right] = \frac{V_3}{T}(T_2 - T_1) \tag{5.38}$$

$$V_A = \frac{V_2 V_3}{V_T}$$

The first input voltage V_1 is compared with the generated triangular wave V_{T1} by the comparator on OA_3. An asymmetrical rectangular waveform V_N is generated at the comparator OA_3 output. From the waveforms shown in Figure 5.12, it is observed that

$$T_3 = \frac{V_T - V_1}{2V_T} T,\ T_4 = \frac{V_T + V_1}{2V_T} T, T = T_3 + T_4 \tag{5.39}$$

This rectangular wave V_N is given as control input to the multiplexer M_2. The multiplexer M_2 connects the output voltage $+V_O$ during T_4 ('by' is connected to 'b') and $-V_O$ during T_3 ('bx' is connected to 'b'). Another rectangular asymmetrical wave V_K with peak to peak values of $\pm V_O$ is generated at the multiplexer M_2 output. The R_7C_3 low pass filter gives the average value of the pulse train V_N and is given as

$$V_B = \frac{1}{T}\left[\int_{O}^{T_4} V_O\, dt + \int_{T_4}^{T_3+T_4} (-V_O)\, dt\right] = \frac{V_O}{T}(T_4 - T_3) \tag{5.40}$$

$$V_B = \frac{V_1 V_O}{V_T}$$

The op-amp OA_4 is configured in a negative closed loop feedback, and a positive dc voltage is ensured in the feedback loop. Hence its inverting terminal voltage must be equal to its non-inverting terminal voltage, i.e.,

$$V_A = V_B \tag{5.41}$$

From equations (5.40) and (5.41),

$$V_O = \frac{V_2 V_3}{V_1} \tag{5.42}$$

5.3 MULTIPLIER CUM DIVIDERS FROM 555 ASTABLE MULTIVIBRATOR

5.3.1 Type I

The circuit diagram of an MCD using the 555 astable is shown in Figure 5.13, and its associated waveforms are shown in Figure 5.14. Refer to the internal diagram of the 555 timer IC shown in Figure 0.1. Initially when we switch on the power supply, the output of the upper comparator CMP_1 will be LOW, i.e., R = 0, and the output of the lower comparator CMP_2 will be HIGH, i.e., S = 1. The flip flop outputs are Q = 1 and Q′ = 0. The timer output at pin 3 will be HIGH, transistor Q_1 is OFF, and hence the discharge pin 7 is at the open position.

The capacitor C_1 is charging toward V_1 through the resistors R_1 and R_2 with a time constant of $(R_1+R_2)C_1$, and its voltage is rising exponentially. When the capacitor voltage is rising above the voltage V_2, the output of the upper comparator CMP_1 becomes HIGH, i.e., R = 1, and the output of the lower comparator CMP_2 becomes LOW, i.e., S = 0. The flip flop outputs are Q = 0 and Q′ = 1. The timer output at pin 3 will be LOW, transistor Q_1 is ON, and hence the discharge pin 7 is at GND potential. Now the capacitor C_1 is discharging to GND potential through the resistor R_2 with a time constant of R_2C. When the capacitor voltage falls below 1/3 V_{CC}, the output

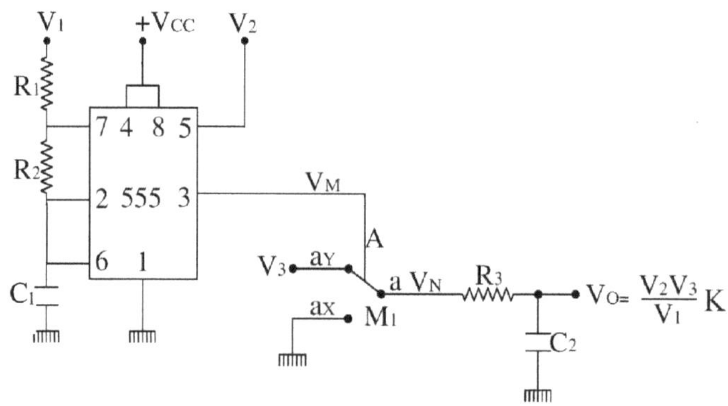

Figure 5.13 555 timer astable as MCD.

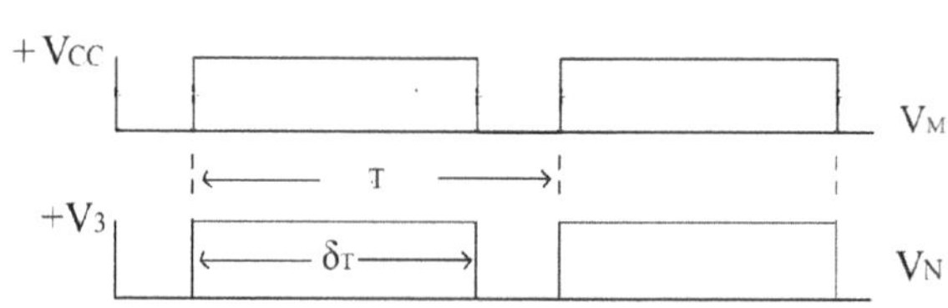

Figure 5.14 Associated waveforms of Figure 5.13.

of the upper comparator CMP_1 becomes LOW, i.e., R = 0, and the output of the lower comparator CMP_2 becomes HIGH, i.e., S = 1. The flip flop outputs are Q = 1 and Q′ = 0. The timer output at pin 3 will be HIGH, transistor Q_1 is OFF, and hence the discharge pin 7 is at the open position.

Now the capacitor starts charging toward V_1, and the cycle therefore repeats to produce periodic pulses at the output pin 3 of the 555 timer.

The ON time δ_T of the 555 timer output V_M is (1) proportional to V_2, which is applied at its pin 5, and (2) inversely proportional to the voltage V_1. During ON time δ_T, V_3 is connected to V_N ('ay' is connected to 'a'). During the OFF time of the waveform V_M, zero voltage is connected to V_N ('ax' is connected to 'a'). Another rectangular waveform V_N with V_3 as peak value is generated at the output of the multiplexer M_1.

$$\delta_T = K\frac{V_2}{V_1}T \tag{5.43}$$

The R_3C_2 low pass filter gives the average value of this pulse train V_N and is given as

$$V_O = \frac{1}{T}\int_0^{\delta_T} V_3 dt = \frac{V_3}{T}\delta_T$$

$$V_O = \frac{V_2V_3}{V_1}K \tag{5.44}$$

where K is a constant value.

5.3.2 Type II Square Wave Referenced MCD

The circuit diagrams of the square wave referenced MCD is shown in Figure 5.15, and its associated waveforms are shown in Figure 5.16. A square waveform V_C is generated by the 555 timer. During the LOW of the square wave, the multiplexer M_1 connects 'ax' to 'a', an integrator, formed of the resistor R_1, capacitor C_1, and op-amp OA_1, integrates the first input voltage $-V_1$. The integrated output will be

$$V_{S1} = -\frac{1}{R_1C_1}\int -V_1 dt = \frac{V_1}{R_1C_1}t \tag{5.45}$$

A positive going ramp Vs_1 is generated at the output of op-amp OA_1. During the HIGH of the square waveform, the multiplexer M_1 connects 'ay' to 'a', and hence the capacitor C_1 is shorted so that op-amp OA_1 output becomes zero. The cycle therefore repeats to provide a semi-saw tooth wave of peak value V_R at the output of op-amp OA_1.

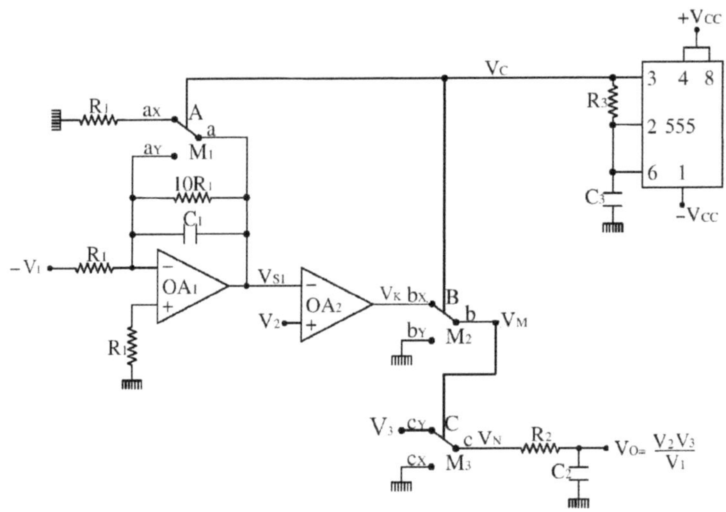

Figure 5.15 Square wave referenced MCD.

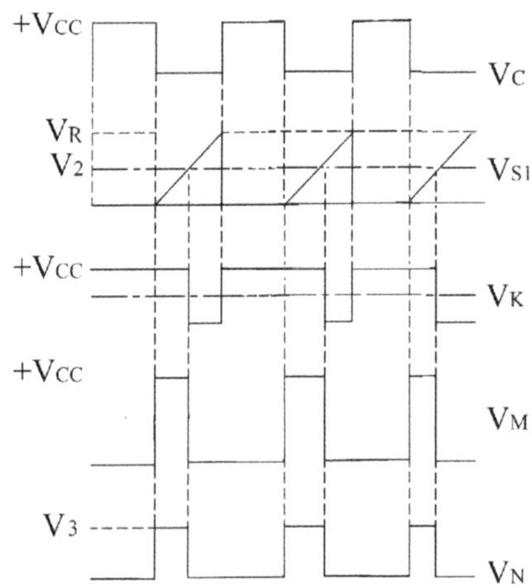

Figure 5.16 Associated waveforms of Figure 5.15.

From the waveforms shown in Figure 5.16 and from equation (5.45), at $t = T/2$, $V_{S1} = V_R$:

$$V_R = \frac{V_1}{R_1C_1}\frac{T}{2}$$

$$T/2 = \frac{V_R}{V_1}R_1C_1 \tag{5.46}$$

The comparator OA_2 compares the semi-saw tooth wave V_{S1} of peak value V_R with the input voltage V_2 and produces a rectangular waveform V_K at its output. The square wave V_C controls the second multiplexer M_2. The multiplexer M_2 connects zero volts during the HIGH of V_C and V_K during the LOW of V_C. Another rectangular waveform V_M is generated at the output of multiplexer M_2. The ON time δ_T of this rectangular waveform V_M is given as

$$\delta_T = \frac{V_2}{V_R}\frac{T}{2} \tag{5.47}$$

The rectangular pulse V_M controls the third multiplexer M_3. When V_M is HIGH, the third input voltage V_3 is connected to the R_2C_2 low pass filter ('cy' is connected to 'c'). When V_M is LOW, zero voltage is connected to the R_2C_2 low pass filter ('cx' is connected to 'c'). Another rectangular pulse V_N with a maximum value of V_3 is generated at the multiplexer M_2 output. The R_2C_2 low pass filter gives the average value of this pulse train V_N and is given as

$$V_O = \frac{1}{T}\int_0^{\delta_T} V_3 dt$$

$$V_O = \frac{V_3}{T}\delta_T \tag{5.48}$$

From equations (5.46)–(5.48),

$$V_O = \frac{V_2 V_3}{V_1}\frac{R_1 C_1}{T} \tag{5.49}$$

Let $T = R_1C_1$.

$$V_O = \frac{V_2 V_3}{V_1} \tag{5.50}$$

5.4 MULTIPLIER CUM DIVIDER FROM 555 MONOSTABLE MULTIVIBRATOR

The circuit diagram of a MCD using the 555 monostable is shown in Figure 5.17, and its associated waveforms are shown in Figure 5.18. Refer to the internal diagram of the 555 timer IC shown in Figure 0.1. Initially when the power supply is switched on, the output of the upper comparator CMP_1 will be LOW, i.e., $R = 0$, and the output of the lower comparator CMP_2 will be HIGH, i.e., $S = 1$. The flip flop outputs are $Q = 1$ and $Q' = 0$. The timer output at pin 3 will be HIGH, transistor Q_1 is OFF, and hence the discharge

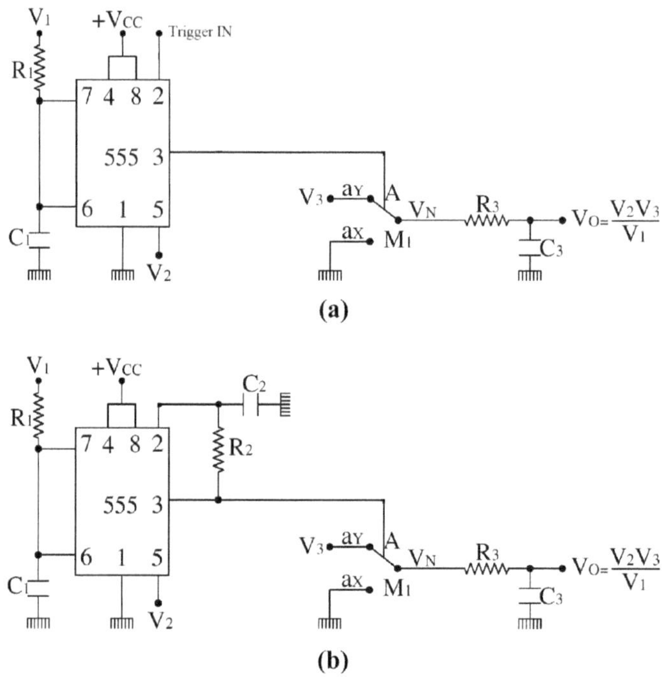

Figure 5.17 (a) 555 monostable as MCD. (b) 555 re-trigger monostable as MCD.

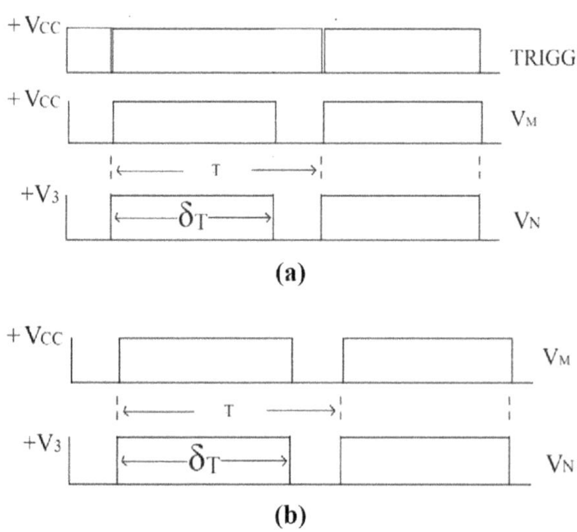

Figure 5.18 (a) Associated waveforms of Figure 5.17(a). (b) Associated waveforms of Figure 5.17(b).

pin 7 is at the open position. The capacitor C_1 is charging toward V_1 through the resistor R_1. The capacitor voltage is rising exponentially, and when it reaches the value of V_2, the output of the upper comparator CMP_1 becomes HIGH, i.e., R = 1, and the output of the lower comparator CMP_2 becomes LOW, i.e., S = 0. The flip flop outputs are Q = 0 and Q′ = 1. The timer output at pin 3 will be LOW, transistor Q_1 is ON, and hence the discharge pin 7 is at GND potential. Now the capacitor C_1 is short circuited, zero voltage exists at pin 6, the output of the upper comparator CMP_1 becomes LOW, i.e., R = 0. A trigger pulse is applied at pin 2, and when the trigger voltage comes down to 1/3 V_{CC}, the output of the lower comparator CMP_2 becomes HIGH, i.e., S = 1. The flip flop outputs are Q = 1 and Q′ = 0. The timer output at pin 3 will be HIGH, transistor Q_1 is OFF, and hence the discharge pin 7 is at the open position.

Now the capacitor C_1 is charging toward V_1, and the sequence therefore repeats for every trigger input pulse.

The ON time of the 555 timer output V_M is (1) proportional to V_2, which is applied at its pin 5, and (2) inversely proportional to the input voltage V_1. During the ON time δ_T, V_3 is connected to the R_3C_3 low pass filter ('ay' is connected to 'a'). During the OFF time of the waveform V_M, zero voltage is connected to the R_3C_3 low pass filter). Another rectangular waveform V_N with V_3 as the peak value is generated at the output of multiplexer M_1. The ON time δ_T of this rectangular waveform V_N is given as

$$\delta_T = K\frac{V_2}{V_1}T \tag{5.51}$$

The R_3C_3 low pass filter gives the average value of this pulse train V_N and is given as

$$V_O = \frac{1}{T}\int_0^{\delta_T} V_3 dt = \frac{V_3}{T}\delta_T$$

$$V_O = \frac{V_2 V_3}{V_1}K \tag{5.52}$$

where K is a constant value.

The MCD using a re-trigger monostable multivibrator is shown in Figure 5.17(b).

Chapter 6

Time Division Multiplier cum Divider—Switching

6.1 SAW TOOTH WAVE BASED MCDS

6.1.1 Saw Tooth Wave Based Double Switching-Averaging Time Division MCD

The circuit diagrams of double switching—averaging time division MCDs are shown in Figure 6.1, and their associated waveforms are shown in Figure 6.2. Figure 6.1(a) shows a series switching MCD, and Figure 6.1(b) shows a shunt switching MCD. A saw tooth wave V_{S1} of peak value V_R and time period T is generated around the 555 timer.

The comparator OA_3 compares the saw tooth wave with the voltage V_Y and produces a first rectangular waveform V_K. The ON time [Figure 6.1(a)] or the OFF time [Figure 6.1(b)] δ_T of V_K is given as

$$\delta_T = \frac{V_Y}{V_R} T \tag{6.1}$$

The rectangular pulse V_K controls the switches S_2 and S_3

- In Figure 6.1(a), when V_K is HIGH, the switch S_2 is closed, and a third input voltage V_3 is connected to the R_3C_2 low pass filter; the switch S_3 is closed, and the first input voltage V_1 is connected to R_4C_3 low pass filter. When V_K is LOW, the switch S_2 is opened, and zero voltage exists on the R_3C_2 low pass filter; the switch S_3 is opened, and zero voltage exists on the R_4C_3 low pass filter.
- In Figure 6.1(b), when V_K is HIGH, the switch S_2 is closed, and zero voltage exists on the R_3C_2 low pass filter; the switch S_3 is closed, and zero voltage exists on the R_4C_3 low pass filter. When V_K is LOW, the switch S_2 is opened, and a third input voltage V_3 is connected to the R_3C_2 low pass filter; the switch S_3 is opened, and the first input voltage V_1 is connected to the R_4C_3 low pass filter.

DOI: 10.1201/9781003362968-6

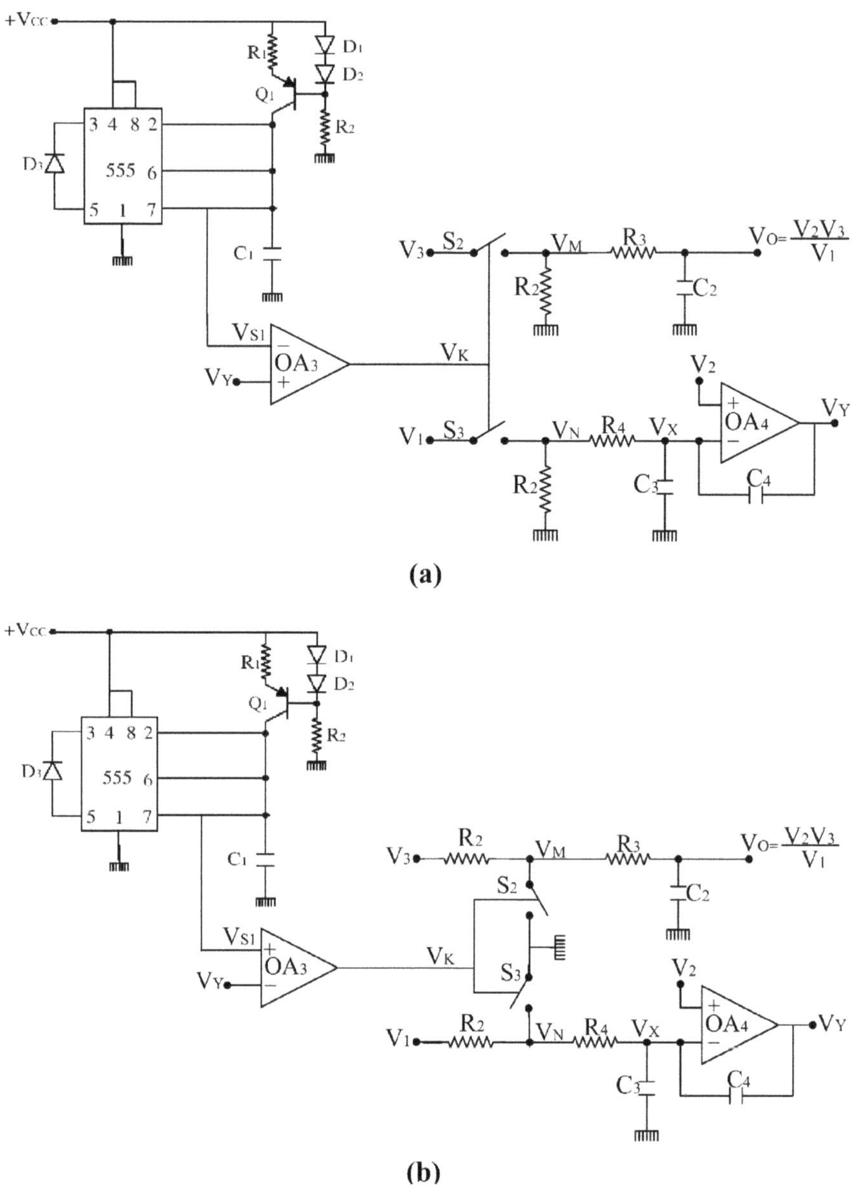

Figure 6.1 (a) Double series switching—averaging time division MCD. (b) D ouble shunt switching—averaging time division MCD.

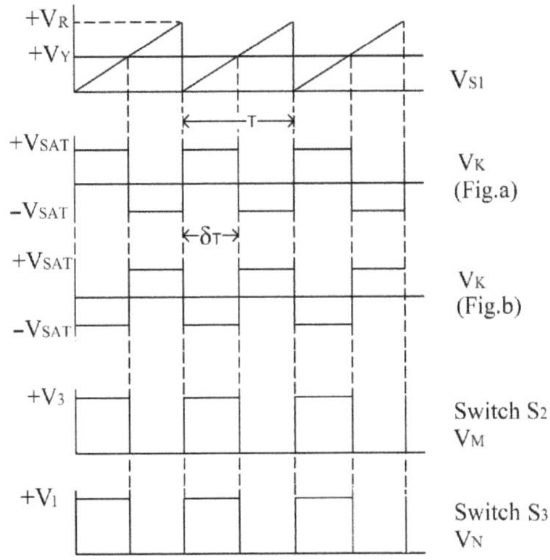

Figure 6.2 Associated waveforms of Figure 6.1.

A second rectangular pulse V_N with a maximum value of V_1 is generated at the switch S_3 output. The R_4C_3 low pass filter gives the average value of this pulse train V_N and is given as

$$V_X = \frac{1}{T}\int_0^{\delta_T} V_1 dt = \frac{V_1}{T}\delta_T$$

$$V_X = \frac{V_1 V_Y}{V_R} \tag{6.2}$$

The op-amp OA_4 is configured in a negative closed loop feedback, and a positive dc voltage is ensured in the feedback loop. Hence its inverting terminal voltage must be equal to its non-inverting terminal voltage.

$$V_2 = V_X \tag{6.3}$$

From equations (6.2) and (6.3)

$$V_Y = \frac{V_2 V_R}{V_1} \tag{6.4}$$

A third rectangular pulse V_M with a maximum value of V_3 is generated at the switch S_2 output. The R_3C_2 low pass filter gives the average value of this pulse train V_M and is given as

$$V_O = \frac{1}{T}\int_0^{\delta_T} V_3 dt = \frac{V_3}{T}\delta_T \qquad (6.5)$$

Equations (6.1) and (6.4) in (6.5) give

$$V_O = \frac{V_2 V_3}{V_1} \qquad (6.6)$$

6.1.2 Saw Tooth Wave Referenced Time Division Multiply-Divide MCD

The circuit diagrams of saw tooth wave based time division multiply--divide MCDs are shown in Figure 6.3, and their associated waveforms are shown in Figure 6.4. Figure 6.3(a) shows a series switching MCD, and Figure 6.3(b) shows a shunt switching MCD. A saw tooth wave V_{S1} of period T is generated around the 555 timer. The comparator OA_2 compares the saw tooth wave with an input voltage V_3 and produces a rectangular waveform V_M.

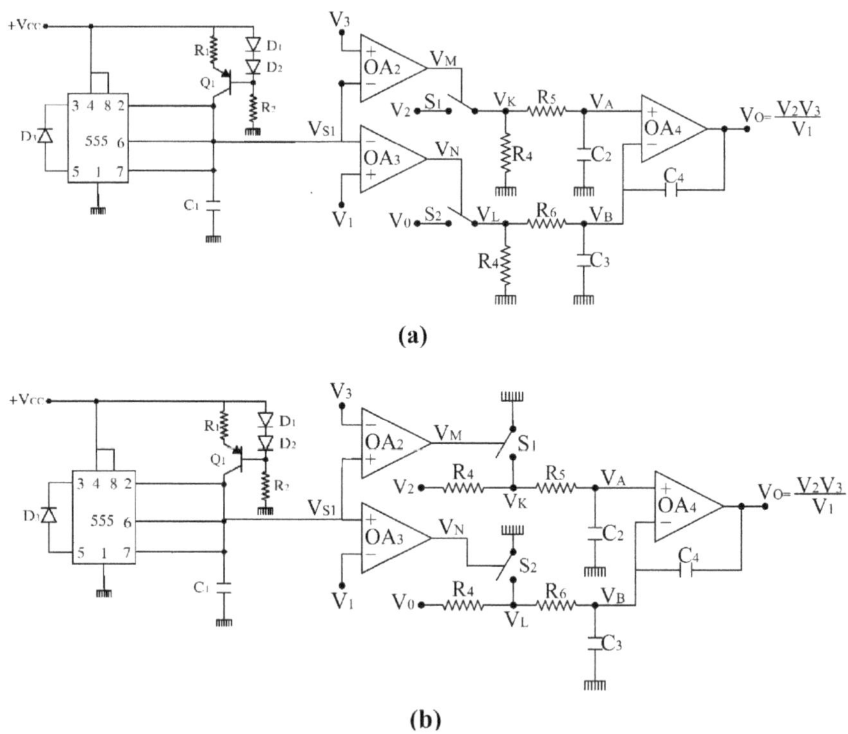

Figure 6.3 (a) Series switching time division multiply-divide MCD. (b) S hunt switching time division multiply-divide MCD.

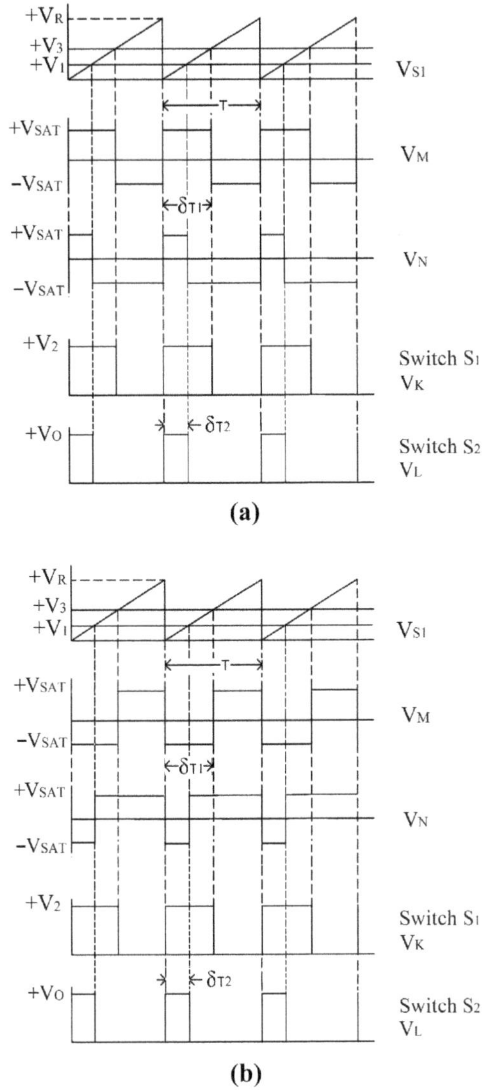

Figure 6.4 (a) Associated waveforms of Figure 6.3(a). (b) Associated waveforms of Figure 6.3(b).

The ON time $\delta_{T1}$of V_M [Figure 6.3(a)] or the OFF time δ_{T1} of V_M [Figure 6.3(b)] is given as

$$\delta_{T1} = \frac{V_3}{V_R} T \tag{6.7}$$

The rectangular pulse V_M controls the first switch S_1.

- In Figure 6.3(a), when V_M is HIGH, the switch S_1 is closed, and another input voltage V_2 is connected to R_5C_2 low pass filter. When V_M is LOW, the switch S_1 is opened, zero voltage is connected to the R_5C_2 low pass filter.
- In Figure 6.3(b), when V_M is HIGH, the switch S_1 is closed, and zero voltage is connected to the R_5C_2 low pass filter. When V_M is LOW, S_1 is opened, and another input voltage V_2 is connected to the R_5C_2 low pass filter.

Another rectangular pulse V_K with maximum value of V_2 is generated at the switch S_1 output. The R_5C_2 low pass filter gives the average value of this pulse train V_K and is given as

$$V_A = \frac{1}{T}\int_0^{\delta_{T1}} V_2 dt = \frac{V_2}{T}\delta_{T1}$$

$$V_A = \frac{V_2 V_3}{V_R} \tag{6.8}$$

The comparator OA_3 compares the saw tooth wave V_{S1} with the first input voltage V_1 and produces a rectangular waveform V_N. The ON time δ_{T2} of V_N, [Figure 6.3(a)] or the OFF time δ_{T2} of V_N [Figure 6.3(b)] is given as

$$\delta_{T2} = \frac{V_1}{V_R} T \tag{6.9}$$

The rectangular pulse V_N controls the second switch S_2.

- In Figure 6.3(a), When V_N is HIGH, the output voltage V_O is connected to the R_6C_3 low pass filter (switch S_2 is closed). When V_N is LOW, zero voltage is connected to the R_6C_3 low pass filter (switch S_2 is opened).
- In Figure 6.3(b), when V_N is HIGH, zero voltage is connected to the R_6C_3 low pass filter (switch S_2 is closed). When V_N is LOW, output voltage V_O is connected to the R_6C_3 low pass filter (switch S_2 is opened).

Another rectangular pulse V_L with maximum value of V_O is generated at the switch S_2 output. The R_6C_3 low pass filter gives the average value of this pulse train V_L and is given as

$$V_B = \frac{1}{T}\int_0^{\delta_{T2}} V_O dt = \frac{V_O}{T}\delta_{T2}$$

$$V_B = \frac{V_1 V_O}{V_R} \tag{6.10}$$

The op-amp OA_4 is configured in a negative closed loop feedback, and a positive dc voltage is ensured in the feedback loop. Hence its inverting terminal voltage must be equal to its non-inverting terminal voltage, i.e.,

$$V_A = V_B \tag{6.11}$$

From equations (6.8) and (6.10),

$$V_O = \frac{V_2 V_3}{V_1} \tag{6.12}$$

6.1.3 Saw Tooth Wave Referenced Time Division Divide-Multiply MCD

The circuit diagrams of saw tooth wave based time division divide-multiply MCDs are shown in Figure 6.5, and their associated waveforms are shown in Figure 6.6. Figure 6.5(a) shows a series switching MCD, and Figure 6.5(b)

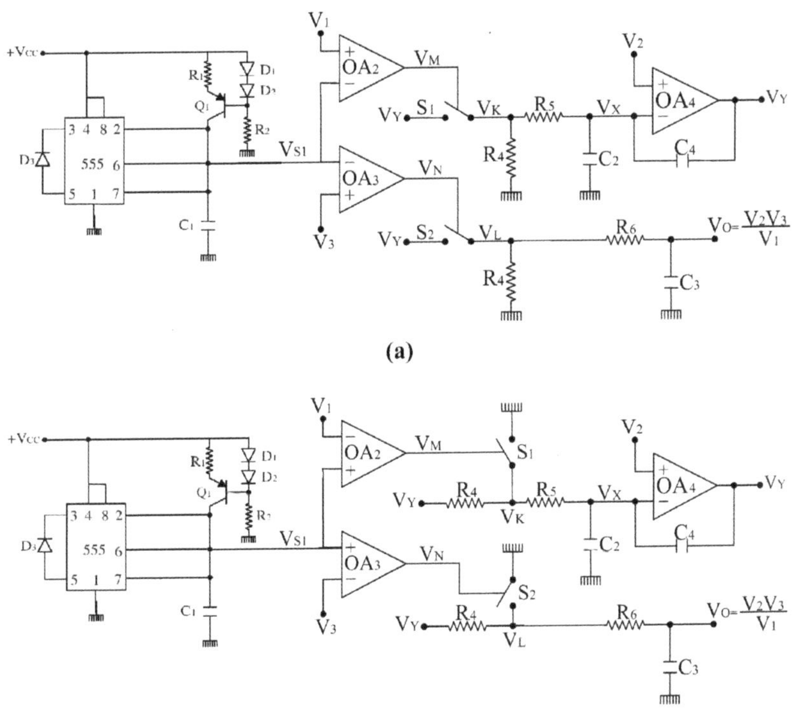

Figure 6.5 (a) Series switching time division divide-multiply MCD. (b) Shunt switching time division divide-multiply MCD.

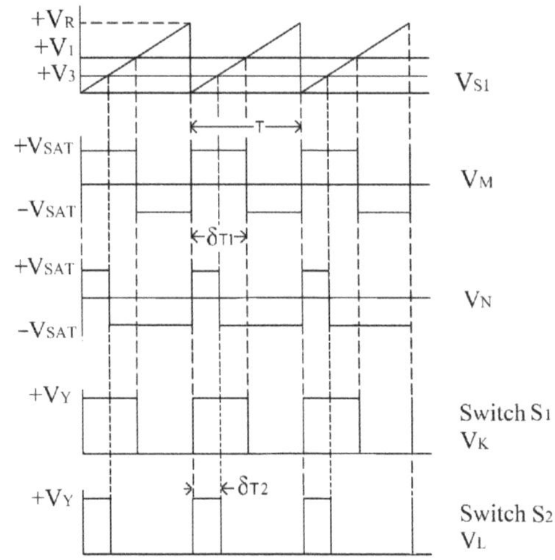

(a)

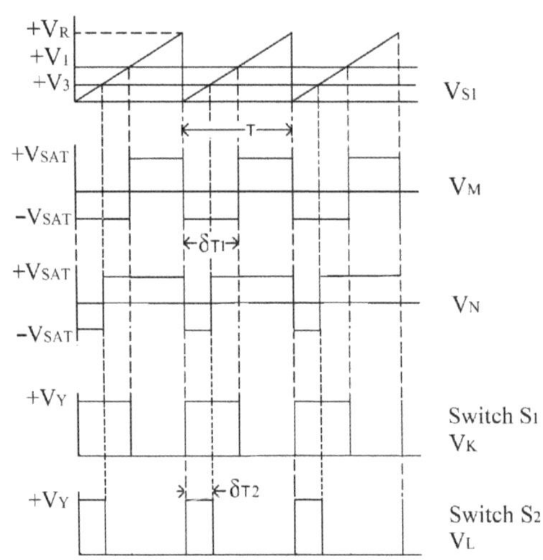

(b)

Figure 6.6 (a) Associated waveforms of Figure 6.5(a). (b) Associated waveforms of Figure 6.5(b).

shows a shunt switching MCD. A saw tooth wave V_{S1} of V_R peak value and time period T is generated around the 555 timer. The comparator OA_2 compares the saw tooth wave with an input voltage V_1 and produces a rectangular waveform V_M. The ON time δ_{T1} of V_M, [Figure 6.7(a)] or the OFF time δ_{T1} of V_M [Figure 6.7(b)] is given as

$$\delta_{T1} = \frac{V_1}{V_R} T \tag{6.13}$$

The rectangular pulse V_M controls the first switch S_1.

- In Figure 6.5(a), when V_M is HIGH, the switch S_1 is closed, and the voltage V_Y is connected to the R_5C_2 low pass filter. When V_M is LOW, the switch S_1 is opened, zero voltage is connected to the R_5C_2 low pass filter.
- In Figure 6.5(b), when V_M is HIGH, the switch S_1 is closed, and zero voltage is connected to the R_5C_2 low pass filter. When V_M is LOW, the switch S_1 is opened, and input voltage V_Y is connected to R_5C_2 low pass filter.

Another rectangular pulse V_K with maximum value of V_Y is generated at the switch S_1 output. The R_5C_2 low pass filter gives the average value of this pulse train V_N and is given as

$$V_X = \frac{1}{T}\int_0^{\delta_{T1}} V_Y dt = \frac{V_Y}{T}\delta_{T1}$$

$$V_X = \frac{V_1 V_Y}{V_R} \tag{6.14}$$

The op-amp OA_4 is configured in a negative closed loop feedback, and a positive dc voltage is ensured in the feedback loop. Hence its inverting terminal voltage must be equal to its non-inverting terminal voltage, i.e.,

$$V_X = V_2 \tag{6.15}$$

From equations (6.14) and (6.15),

$$V_Y = \frac{V_2 V_R}{V_1} \tag{6.16}$$

The comparator OA_3 compares the saw tooth wave V_{S1} with the third input voltage V_3 and produces a rectangular waveform V_N. The ON time δ_{T2} of V_N [Figure 6.5(a)] or the OFF time δ_{T2} of V_N [Figure 6.5(b)] is given as

$$\delta_{T2} = \frac{V_3}{V_R} T \tag{6.17}$$

The rectangular pulse V_N controls the second switch S_2.

- In Figure 6.5(a), when V_N is HIGH, the switch S_2 is closed, the voltage V_Y is connected to the R_6C_3 low pass filter. When V_N is LOW, the switch S_2 is opened, zero voltage is connected to the R_6C_3 low pass filter

- In Figure 6.5(b), when V_N is HIGH, the switch S_2 is closed, zero voltage is connected to the R_6C_3 low pass filter. When V_N is LOW, the switch S_2 is opened, and the voltage V_Y is connected to the R_6C_3 low pass filter,

Another rectangular pulse V_L with maximum value of V_Y is generated at the switch S_2 output. The R_6C_3 low pass filter gives the average value of this pulse train V_L and is given as

$$V_O = \frac{1}{T}\int_0^{\delta_{T2}} V_Y dt = \frac{V_Y}{T}\delta_{T2} \tag{6.18}$$

$$V_O = \frac{V_3 V_Y}{V_R} \tag{6.19}$$

Equation (6.16) in (6.19) gives

$$V_O = \frac{V_2 V_3}{V_1} \tag{6.20}$$

6.2 TRIANGULAR WAVE BASED MCDS

6.2.1 Time Division MCD

The circuit diagrams of triangular wave referenced time division MCDs are shown in Figure 6.7, and their associated waveforms are shown in Figure 6.8. Figure 6.7(a) shows a series switching MCD, and Figure 6.7(b) shows a shunt switching MCD. The output of the 555 timer circuit is the triangular wave V_{T1} with $\pm V_T$ peak values and time period T.

The comparator OA_2 compares this triangular wave V_{T1} with the voltage V_Y and produces the asymmetrical rectangular wave V_K. From Figure 6.8, it is observed that

$$T_1 = \frac{V_T - V_Y}{2V_T}T, T_2 = \frac{V_T + V_Y}{2V_T}T, T = T_1 + T_2 \tag{6.21}$$

The rectangular wave V_K controls switches S_1 and S_2. During the ON time T_2 of this rectangular wave V_K:

- In Figure 6.7(a), the switch S_1 is closed, the op-amp OA_5 along with the resistor R_6 will work as a non-inverting amplifier, and $+V_3$ will exist on its output ($V_M = +V_3$). The switch S_2 is closed, the op-amp OA_3 along with the resistor R_7 will work as a non-inverting amplifier, and $+V_1$ will exist on its output ($V_N = +V_1$).

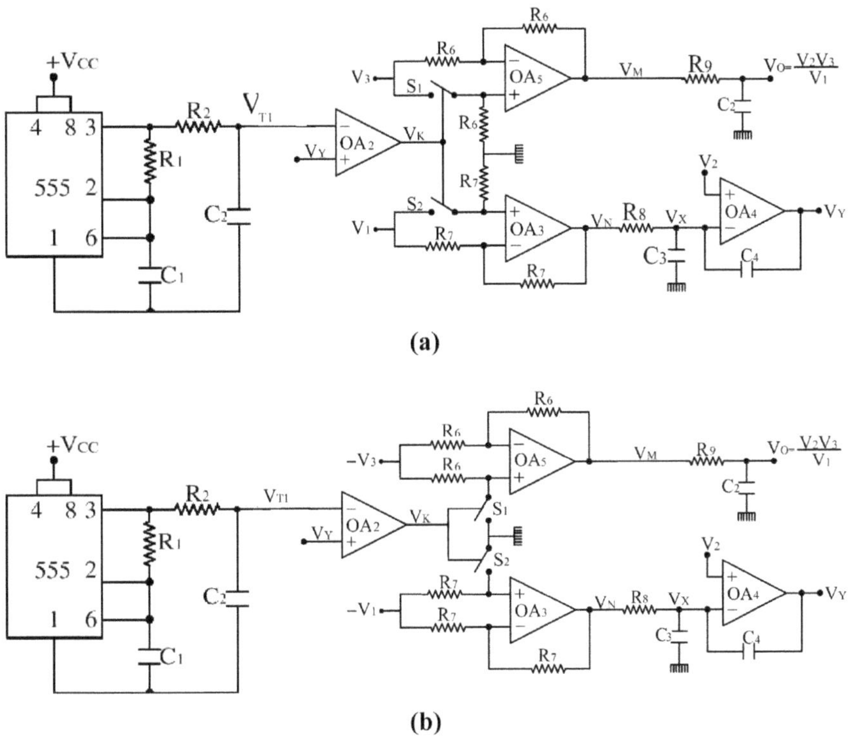

Figure 6.7 (a) Series switching time division MCD. (b) Shunt switching time division MCD.

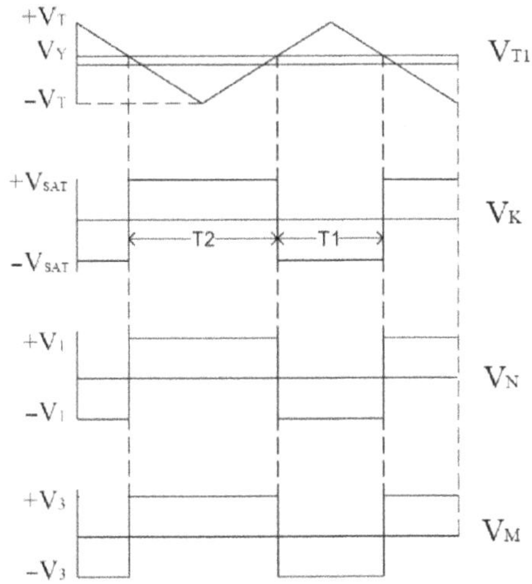

Figure 6.8 Associated wave forms of Figure 6.7.

- In Figure 6.7(b), the switch S_1 is closed, the op-amp OA_5 along with the resistor R_6 will work as an inverting amplifier, and $+V_3$ will exist on its output ($V_M = +V_3$). The switch S_2 is closed, the op-amp OA_3 along with the resistor R_7 will work as an inverting amplifier, and $+V_1$ will exist on its output ($V_N = +V_1$).

During the OFF time T_1 of this rectangular wave V_K:

- In Figure 6.7(a), the switch S_1 is opened, the op-amp OA_5 along with the resistor R_6 will work as an inverting amplifier, and $-V_3$ will exist on its output ($V_M = -V_3$). The switch S_2 is opened, the op-amp OA_3 along with the resistor R_7 will work as an inverting amplifier, and $-V_1$ will exist on its output ($V_N = -V_1$).
- In Figure 6.7(b), the switch S_1 is opened, the op-amp OA_5 along with the resistor R_6 will work as a non-inverting amplifier, and $-V_3$ will exist on its output ($V_M = -V_3$). The switch S_2 is closed, the op-amp OA_3 along with the resistor R_7 will work as a non-inverting amplifier, and $-V_1$ will exist on its output ($V_N = -V_1$).

Two asymmetrical rectangular waveforms (1) V_M with $\pm V_3$ peak to peak values at the output of op-amp OA_5 and (2) V_N with $\pm V_1$ peak to peak values at the output of op-amp OA_3, are generated. The R_8C_3 low pass filter gives the average value of V_N and is given as

$$V_X = \frac{1}{T}\left[\int_{O}^{T_2} V_1\,dt + \int_{T_2}^{T_1+T_2} (-V_1)\,dt\right] = \frac{V_1}{T}[T_2 - T_1] \tag{6.22}$$

$$V_X = \frac{V_1 V_Y}{V_T}$$

The op-amp OA_4 is configured in a negative closed loop feedback, and a positive dc voltage is ensured in the feedback loop. Hence its inverting terminal voltage must be equal to its non-inverting terminal voltage, i.e.,

$$V_X = V_2 \tag{6.23}$$

From equations (6.22) and (6.23),

$$V_Y = \frac{V_2 V_T}{V_1} \tag{6.24}$$

The R_9C_2 low pass filter gives the average value of the rectangular waveform V_M and is given as

$$V_O = \frac{1}{T}\left[\int_{O}^{T_2} V_3\,dt + \int_{T_2}^{T_1+T_2} (-V_3)\,dt\right] = \frac{V_3}{T}[T_2 - T_1] \tag{6.25}$$

$$V_O = \frac{V_3 V_Y}{V_T}$$

Equation (6.24) in (6.25) gives

$$V_O = \frac{V_2 V_3}{V_1} \tag{6.26}$$

6.2.2 Divide-Multiply Time Division MCD

The circuit diagrams of a triangular wave based divide-multiply time division MCD are shown in Figure 6.9, and their associated waveforms are shown in Figure 6.10. A triangular wave V_{T1} with $\pm V_T$ peak to peak value is generated around the 555 timer.

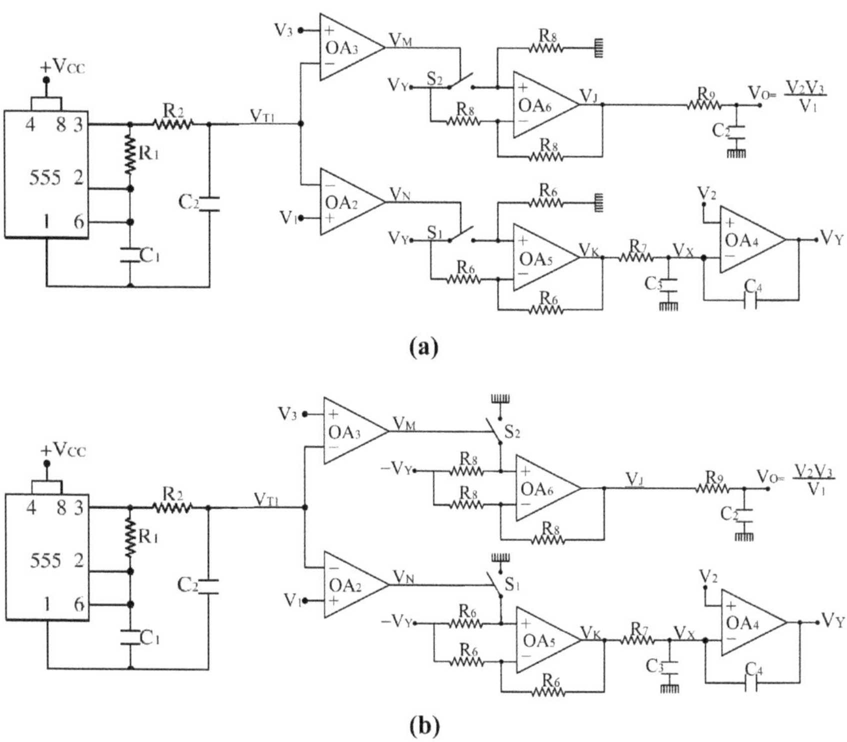

Figure 6.9 (a) Series switching divide-multiply time division MCD. (b) Shunt switching divide-multiply time division MCD.

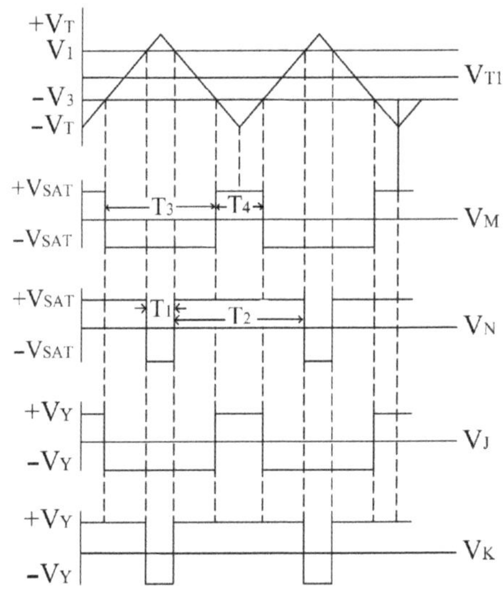

Figure 6.10 Associated waveforms of Figure 6.9.

The first input voltage V_1 is compared with the generated triangular wave V_{T1} by the comparator on OA_2. An asymmetrical rectangular waveform V_N is generated at the comparator OA_2 output. From the waveforms shown in Figure 6.10, it is observed that

$$T_1 = \frac{V_T - V_1}{2V_T} T, T_2 = \frac{V_T + V_1}{2V_T} T, T = T_1 + T_2 \tag{6.27}$$

This rectangular wave V_N is given as the control input to the switch S_1. During the ON time T_2 of this rectangular wave V_N:

- In Figure 6.9(a), the switch S_1 is closed, the op-amp OA_5 along with the resistor R_6 will work as a non-inverting amplifier, and $+V_Y$ will exist on its output ($V_K = +V_Y$).
- In Figure 6.9(b), the switch S_1 is closed, the op-amp OA_5 along with the resistor R_6 will work as an inverting amplifier, and $+V_Y$ will exist on its output ($V_K = +V_Y$).

During the OFF time T_3 of this rectangular wave V_N:

- In Figure 6.9(a), the switch S_1 is opened, the op-amp OA_5 along with the resistor R_6 will work as an inverting amplifier, and $-V_Y$ will exist on its output ($V_K = -V_Y$).

- In Figure 6.9(b), the switch S_1 is opened, the op-amp OA_5 along with the resistor R_6 will work as a non-inverting amplifier, and $-V_Y$ will exist on its output ($V_K = -V_Y$).

Another rectangular asymmetrical wave V_K with peak to peak values of $\pm V_Y$ is generated at the output of op-amp OA_5. The R_7C_3 low pass filter gives the average value of the pulse train V_K and is given as

$$V_X = \frac{1}{T}\left[\int_{O}^{T_2} V_Y \, dt + \int_{T_2}^{T_1+T_2} (-V_Y)dt\right] = \frac{V_Y}{T}(T_2 - T_1)$$

$$V_X = \frac{V_1 V_Y}{V_T} \tag{6.28}$$

The op-amp OA_4 is configured in a negative closed loop feedback, and a positive dc voltage is ensured in the feedback loop. Hence its inverting terminal voltage must be equal to its non-inverting terminal voltage, i.e.,

$$V_2 = V_X \tag{6.29}$$

From equations (6.28) and (6.29),

$$V_Y = \frac{V_2 V_T}{V_1} \tag{6.30}$$

The third input voltage V_3 is compared with the generated triangular wave V_{T1} by the comparator on OA_3. An asymmetrical rectangular waveform V_M is generated at the comparator OA_3 output. From the waveforms shown in Figure 6.14, it is observed that

$$T_3 = \frac{V_T - V_3}{2V_T} T T_4 = \frac{V_T + V_3}{2V_T} T T = T_3 + T_4 \tag{6.31}$$

This rectangular wave V_M is given as the control input to the switch S_2.

During the ON time T_4 of this rectangular wave V_M:

- In Figure 6.9(a), the switch S_2 is closed, the op-amp OA_6 along with the resistor R_8 will work as a non-inverting amplifier, and $+V_Y$ will exist on its output ($V_J = +V_Y$).
- In Figure 6.9(b), the switch S_2 is closed, the op-amp OA_6 along with the resistor R_8 will work as an inverting amplifier, and $+V_Y$ will exist on its output ($V_J = +V_Y$).

During the OFF time T_1 of this rectangular wave V_M:

- In Figure 6.9(a), the switch S_2 is opened, the op-amp OA_6 along with the resistor R_8 will work as an inverting amplifier, and $-V_Y$ will exist on its output ($V_J = -V_Y$).
- In Figure 6.9(b), the switch S_2 is opened, the op-amp OA_6 along with the resistor R_8 will work as a non-inverting amplifier, and $-V_Y$ will exist on its output ($V_J = -V_Y$).

Another rectangular asymmetrical wave V_J with peak to peak values of $\pm V_Y$ is generated at the output of op-amp OA_6. The R_9C_2 low pass filter gives the average value of the pulse train V_J and is given as

$$V_O = \frac{1}{T}\left[\int_O^{T_4} V_Y\,dt + \int_{T_4}^{T_3+T_4} (-V_Y)\,dt\right] = \frac{V_Y}{T}(T_4 - T_3) \tag{6.32}$$

$$V_O = \frac{V_Y V_3}{V_T} \tag{6.33}$$

Equation (6.30) in (6.33) gives

$$V_O = \frac{V_2 V_3}{V_1} \tag{6.34}$$

6.2.3 Multiply-Divide Time Division MCD

The circuit diagrams of triangular wave based multiply-divide time division MCDs are shown in Figure 6.11, and their associated waveforms are shown in Figure 6.12. Figure 6.11(a) shows a series switching MCD, and Figure 6.11(b) shows a shunt switching MCD. A triangular wave V_{T1} with $\pm V_T$ peak to peak value is generated around the 555 timer.

The second input voltage V_2 is compared with the generated triangular wave V_{T1} by the comparator OA_2. An asymmetrical rectangular waveform V_M is generated at the comparator OA_2 output. From the waveforms shown in Figure 6.12, it is observed that

$$T_1 = \frac{V_T - V_2}{2V_T}T, \quad T_2 = \frac{V_T + V_2}{2V_T}T, \quad T = T_1 + T_2 \tag{6.35}$$

This rectangular wave V_M is given as control input to the switch S_1.

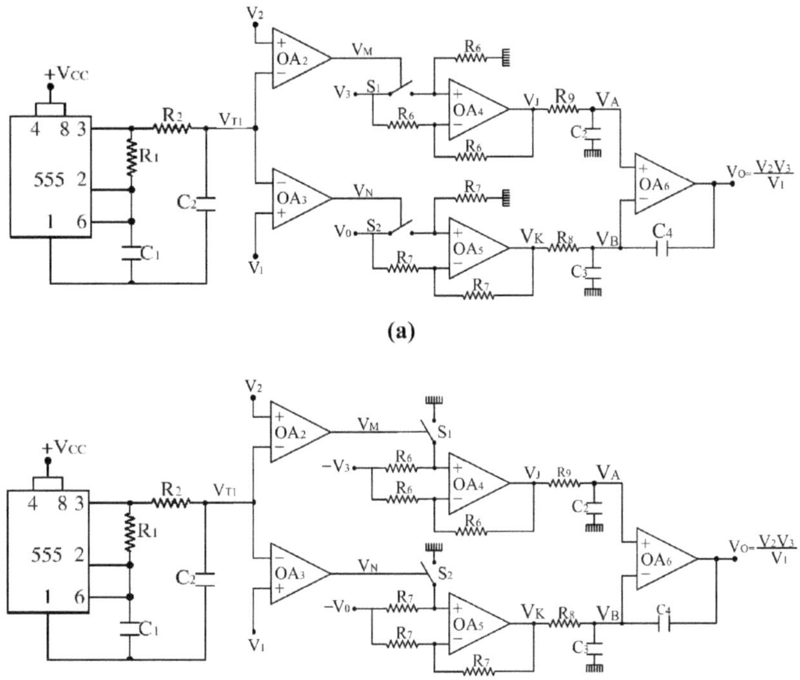

Figure 6.11 (a) Series switching multiply-divide time division MCD. (b) Shunt switching multiply—divide time division MCD.

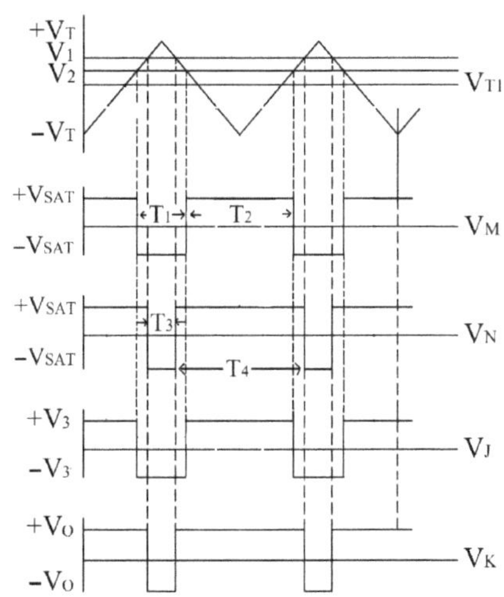

Figure 6.12 Associated waveforms of Figure 6.11.

During the ON time T_2 of this rectangular wave V_M:

- In Figure 6.11(a), the switch S_1 is closed, the op-amp OA_4 along with the resistor R_6 will work as a non-inverting amplifier, and $+V_3$ will exist on its output ($V_J = +V_3$).
- In Figure 6.11(b), the switch S_1 is closed, the op-amp OA_4 along with the resistor R_6 will work as an inverting amplifier, and $+V_3$ will exist on its output ($V_J = +V_3$).

During the OFF time T_1 of this rectangular wave V_M:

- In Figure 6.11(a), the switch S_1 is opened, the op-amp OA_4 along with the resistor R_6 will work as an inverting amplifier, and $-V_3$ will exist on its output ($V_J = -V_3$).
- In Figure 6.11(b), the switch S_1 is opened, the op-amp OA_4 along with the resistor R_6 will work as a non-inverting amplifier, and $-V_3$ will exist on its output ($V_J = -V_3$).

Another rectangular asymmetrical wave V_J with peak to peak values of $\pm V_3$ is generated at the output of op-amp OA_4. The R_9C_2 low pass filter gives the average value of the pulse train V_J and is given as

$$V_A = \frac{1}{T}\left[\int_{O}^{T_2} V_3\,dt + \int_{T_2}^{T_1+T_2} (-V_3)\,dt\right] = \frac{V_3}{T}(T_2 - T_1) \tag{6.36}$$

$$V_A = \frac{V_2 V_3}{V_T}$$

The first input voltage V_1 is compared with the generated triangular wave V_{T1} by the comparator on OA_3. An asymmetrical rectangular waveform V_N is generated at the comparator OA_3 output. From the waveforms shown in Figure 6.17, it is observed that

$$T_3 = \frac{V_T - V_1}{2V_T}T, \quad T_4 = \frac{V_T + V_1}{2V_T}T, \quad T = T_3 + T_4 \tag{6.37}$$

This rectangular wave V_N is given as control input to the switch S_2.

During the ON time T_4 of this rectangular wave V_N:

- In Figure 6.11(a), the switch S_2 is closed, the op-amp OA_5 along with the resistor R_7 will work as a non-inverting amplifier, and $+V_O$ will exist on its output ($V_K = +V_O$).
- In Figure 6.11(b), the switch S_2 is closed, the op-amp OA_5 along with the resistor R_7 will work as an inverting amplifier, and $+V_O$ will exist on its output ($V_K = +V_O$).

During the OFF time T_3 of this rectangular wave V_N:

- In Figure 6.11(a), the switch S_2 is opened, the op-amp OA_5 along with the resistor R_7 will work as an inverting amplifier, and $-V_O$ will exist on its output ($V_K = -V_O$).
- In Figure 6.11(b), the switch S_2 is opened, the op-amp OA_5 along with the resistor R_7 will work as a non-inverting amplifier, and $-V_O$ will exist on its output ($V_K = -V_O$).

Another rectangular asymmetrical wave V_K with peak to peak values of $\pm V_O$ is generated at the output of the op-amp OA_5. The R_8C_3 low pass filter gives the average value of the pulse train V_K and is given as

$$V_B = \frac{1}{T}\left[\int_0^{T_4} V_O\,dt + \int_{T_4}^{T_3+T_4} (-V_O)\,dt\right] = \frac{V_O}{T}(T_4 - T_3) \tag{6.38}$$

$$V_B = \frac{V_1 V_O}{V_T}$$

The op-amp OA_6 is configured in a negative closed loop feedback, and a positive dc voltage is ensured in the feedback loop. Hence its inverting terminal voltage must be equal to its non-inverting terminal voltage, i.e.,

$$V_A = V_B \tag{6.39}$$

From equations (6.36) and (6.38),

$$V_O = \frac{V_2 V_3}{V_1} \tag{6.40}$$

6.3 MULTIPLIER CUM DIVIDER FROM 555 ASTABLE MULTIVIBRATOR

6.3.1 Multiplier cum Divider from 555 Astable Multivibrator—Type I

The circuit diagram of an MCD using the 555 astable is shown in Figure 6.13, and its associated waveforms are shown in Figure 6.14. Refer to the internal diagram of the 555 timer IC shown in Figure 0.1. Initially when we switch on the power supply, the output of the upper comparator CMP_1 will be LOW, i.e., R = 0, and the output of the lower comparator CMP_2 will be HIGH, i.e., S = 1. The flip flop outputs are Q = 1 and Q′ = 0. The timer output at pin 3 will be HIGH, transistor Q_1 is OFF, and hence the discharge pin 7 is at the open position.

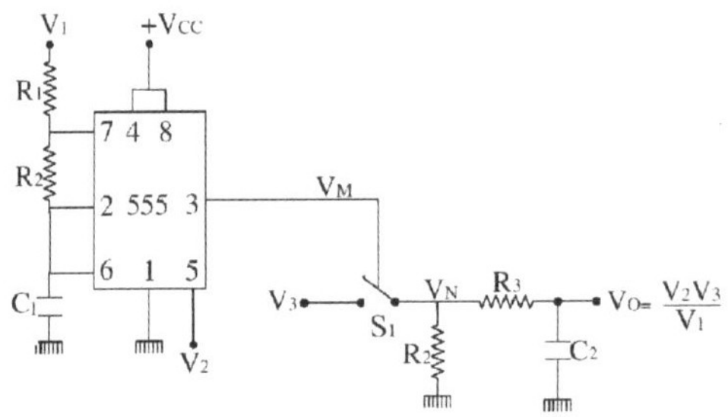

Figure 6.13 555 timer astable as MCD.

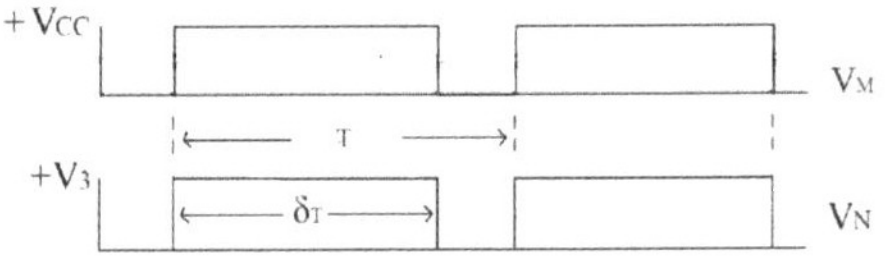

Figure 6.14 Associated waveforms of Figure 6.13.

The capacitor C_1 is charging toward V_1 through the resistors R_1 and R_2 with a time constant of $(R_1+R_2)C_1$, and its voltage is rising exponentially. When the capacitor voltage is rising above the voltage V_2, the output of the upper comparator CMP_1 becomes HIGH, i.e., R = 1, and the output of the lower comparator CMP_2 becomes LOW, i.e., S = 0. The flip flop outputs are Q = 0 and Q′ = 1. The timer output at pin 3 will be LOW, transistor Q_1 is ON, and hence the discharge pin 7 is at GND potential. Now the capacitor C_1 is discharging to GND potential through the resistor R_2 with a time constant of R_2C. When the capacitor voltage falls below 1/3 V_{CC}, the output of the upper comparator CMP_1 becomes LOW, i.e., R = 0, and the output of the lower comparator CMP_2 becomes HIGH, i.e., S = 1. The flip flop outputs are Q = 1 and Q′ = 0. The timer output at pin 3 will be HIGH, transistor Q_1 is OFF, and hence the discharge pin 7 is at the open position.

Now the capacitor starts charging toward V_1, and the cycle therefore repeats to produce periodic pulses at the output pin 3 of the 555 timer.

The ON time δ_T of the 555 timer output V_M is (1) proportional to V_2, which is applied at its pin 5 and (2) inversely proportional to the voltage V_1. During the ON time δ_T, the switch S_1 is closed, its collector voltage V_3 exists at the output of switch S_1. During the OFF time of the waveform V_M, the switch S_1 is opened, zero voltage exists at the output of switch S_1. Another rectangular waveform V_N with V_3 as the peak value is generated at the output of the switch S_1.

The ON time of the rectangular wave V_M is given as

$$\delta_T = K\frac{V_2}{V_1}T \tag{6.41}$$

The R_3C_2 low pass filter gives the average value of this pulse train V_N and is given as

$$V_O = \frac{1}{T}\int_0^{\delta_T} V_3 dt = \frac{V_3}{T}\delta_T$$

$$V_O = \frac{V_2 V_3}{V_1}K \tag{6.42}$$

where K is a constant value.

6.3.2 Square Wave Referenced MCD

The circuit diagrams of a square wave referenced MCD is shown in Figure 6.15, and its associated waveforms are shown in Figure 6.16. During the LOW of the square wave, the switch S_1 is opened, an integrator formed by resistor R_1, capacitor C_1, and op-amp OA_1 integrates the reference voltage $-V_1$. The integrated output will be

$$V_{S1} = -\frac{1}{R_1C_1}\int -V_1 dt = \frac{V_1}{R_1C_1}t \tag{6.43}$$

A positive going ramp Vs_1 is generated at the output of op-amp OA_1. During the HIGH of square waveform, the switch S_1 is closed, and hence the capacitor C_1 is shorted so that the op-amp OA_1 output becomes zero. The cycle therefore repeats to provide a semi-saw tooth wave of peak value V_R at the output of op-amp OA_1.

From the waveforms shown in Figure 6.16 and from equation (6.43), at $t = T/2$, $V_{S1} = V_R$.

$$V_R = \frac{V_1}{R_1C_1}\frac{T}{2}$$

$$T/2 = \frac{V_R}{V_1}R_1C_1 \tag{6.44}$$

The comparator OA_2 compares the semi-saw tooth wave V_{S1} of peak value V_R with the input voltage V_1 and produces a rectangular waveform V_K at its output. The square wave V_C controls the second switch S_2. The switch S_2

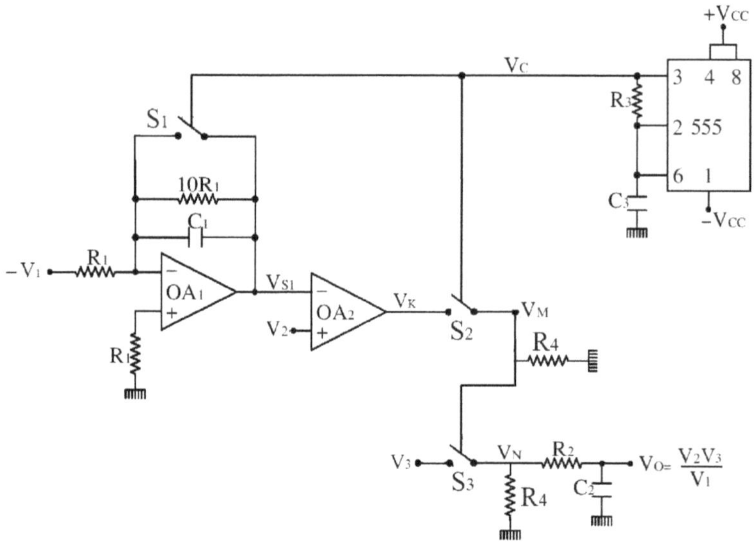

Figure 6.15 Square wave referenced MCD.

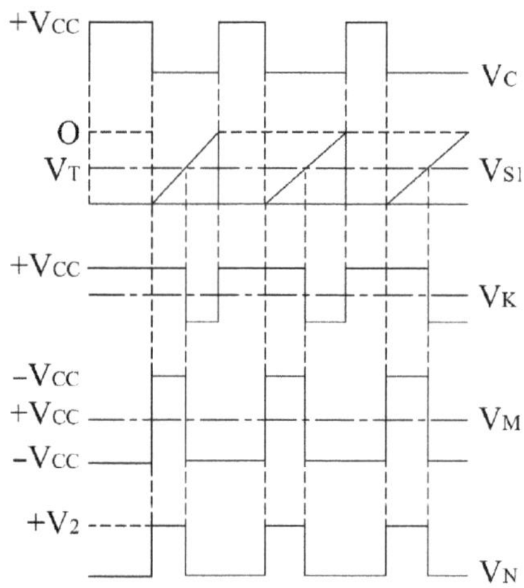

Figure 6.16 Associated waveforms of Figure 6.15.

is closed during the HIGH time of V_C. Another rectangular waveform V_M is generated at the output of switch S_2. The ON time δ_T of this rectangular waveform V_M is given as

$$\delta_T = \frac{V_2}{V_R}\frac{T}{2} \tag{6.45}$$

The rectangular pulse V_M controls the third switch S_3. The switch S_3 is closed during the HIGH time of V_M. Another rectangular pulse V_N with maximum value of V_3 is generated at the switch S_3 output. The R_2C_2 low pass filter gives the average value of this pulse train V_N and is given as

$$V_O = \frac{1}{T}\int_0^{\delta_T} V_3 dt \tag{6.46}$$

$$V_O = \frac{V_3}{T}\delta_T \tag{6.47}$$

From equations (6.45)–(6.47),

$$V_O = \frac{V_2 V_3}{V_1}\frac{R_1C_1}{T}$$

Let $T = R_1C_1$.

$$V_O = \frac{V_2 V_3}{V_1} \tag{6.48}$$

6.4 MULTIPLIER CUM DIVIDER FROM 555 MONOSTABLE MULTIVIBRATOR

The circuit diagram of MCD using the 555 monostable is shown in Figure 6.17, and its associated waveforms are shown in Figure 6.18. Refer to the internal diagram of the 555 timer IC shown in Figure 0.1. Initially when the power supply is switched on, the output of the upper comparator CMP_1 will be LOW, i.e., R = 0, and the output of the lower comparator CMP_2 will be HIGH, i.e., S = 1. The flip flop outputs are Q = 1 and Q′ = 0. The timer output at pin 3 will be HIGH, transistor Q_1 is OFF, and hence the discharge pin 7 is at the open position. The capacitor C_1 is charging toward V_1 through the resistor R_1. The capacitor voltage is rising exponentially, and when it reaches the value of V_2, the output of the upper comparator CMP_1 becomes HIGH, i.e., R = 1, and the output of the lower comparator CMP_2 becomes LOW, i.e., S = 0. The flip flop outputs are Q = 0 and Q′ = 1. The timer output at pin 3 will be LOW, transistor Q_1 is ON, and hence the discharge pin 7 is at GND potential. Now the capacitor C_1 is short circuited, zero voltage exists at pin 6, and the output of the upper comparator CMP_1 becomes LOW, i.e., R = 0. A trigger pulse is applied at pin 2, and when the trigger voltage comes down to 1/3 V_{CC}, the output of the lower comparator CMP_2 becomes HIGH, i.e., S = 1. The flip flop outputs are Q = 1 and Q′ = 0. The timer output at pin 3 will be HIGH, transistor Q_1 is OFF, and hence the discharge pin 7 is at the open position.

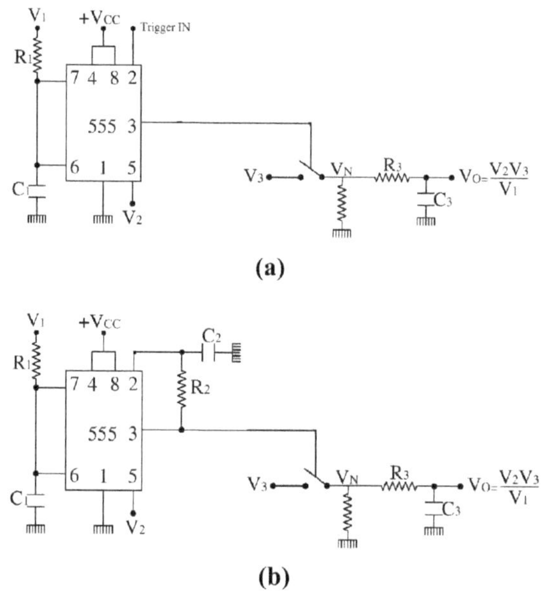

Figure 6.17 (a) 555 monostable as MCD. (b) 555 re-trigger monostable as MCD.

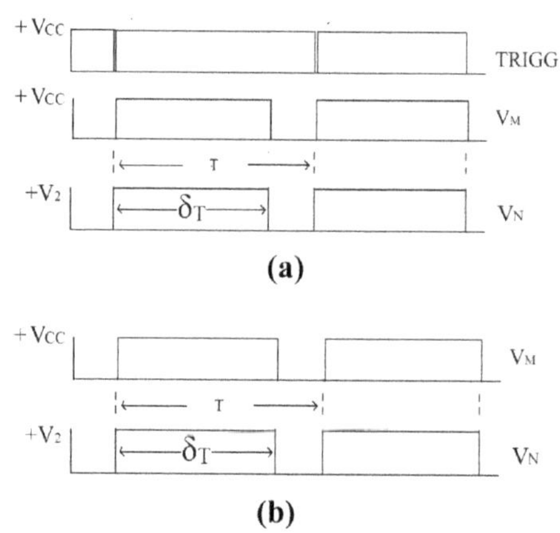

Figure 6.18 (a) Associated waveforms of Figure 6.17(a). (b) Associated waveforms of Figure 6.17(b).

Now the capacitor C_1 is charging toward V_1, and the sequence therefore repeats for every trigger input pulse.

The ON time of the 555 timer output V_M is (1) proportional to V_2, which is applied at its pin 5 and (2) inversely proportional to the input voltage V_1.

During the ON time δ_T, the switch S_1 is closed, and its collector voltage V_3 exists at its emitter terminal. During the OFF time of the waveform V_M, the switch S_1 is opened, zero voltage exists on its emitter terminal. Another rectangular waveform V_N with V_3 as the peak value is generated at the emitter of transistor Q_3. The ON time δ_T of this rectangular waveform V_N is given as

$$\delta_T = K\frac{V_2}{V_1}T \tag{6.49}$$

The R_3C_3 low pass filter gives the average value of this pulse train V_N and is given as

$$V_O = \frac{1}{T}\int_0^{\delta_T} V_3 dt = \frac{V_3}{T}\delta_T$$

$$V_O = \frac{V_2 V_3}{V_1}K \tag{6.50}$$

where K is a constant value.

The MCD using a re-trigger monostable multivibrator is shown in Figure 6.17(b).

Chapter 7

Peak Responding Multiplier cum Dividers—Multiplexing

7.1 DOUBLE SINGLE SLOPE PEAK RESPONDING MCDs

7.1.1 Double Single Slope Peak Responding MCDs—Type I

The circuit diagrams of double single slope peak responding MCDs are shown in Figure 7.1, and their associated waveforms are shown in Figure 7.2. Figure 7.1(a) shows a double single slope peak detecting MCD, and Figure 7.1(b) shows a double single slope peak sampling MCD. Initially the 555 timer output is HIGH, the multiplexer M_1 connects 'ay' to 'a', an integrator formed by the resistor R_1, the capacitor C_1, and the op-amp OA_1 integrates the first input voltage $-V_1$. The integrated output will be

$$V_{S1} = -\frac{1}{R_1C_1}\int -V_1 dt = \frac{V_1}{R_1C_1}t \tag{7.1}$$

A positive going ramp Vs_1 is generated at the output of the op-amp OA_1. When the output of OA_1 reaches the voltage level of V_2, the 555 timer output becomes LOW. The multiplexer M_1 connects 'ax' to 'a', and hence the capacitor C_1 is shorted so that op-amp OA_1 output becomes zero. Then the 555 timer output goes to HIGH, the multiplexer M_1 connects 'ay' to 'a', and the integrator composed by R_1, C_1, and op-amp OA_1 integrates the input voltage $-V_1$, and the cycle therefore repeats to provide (1) a saw tooth wave V_{S1} of peak value V_2 at the output of op-amp OA_1 and (2) a short pulse waveform V_C at the output of the 555 timer. The short pulse V_C also controls the multiplexer M_2. During the short LOW time of V_C, the multiplexer M_2 connects 'bx' to 'b', the capacitor C_2 is short circuited so that the op-amp OA_2 output is zero volts. During the HIGH time of V_C, the multiplexer M_2 connects 'by' to 'b', and the integrator formed by the resistor R_2, capacitor C_2, and op-amp OA_2 integrates its input voltage $-V_3$. Its output is given as

$$V_{S2} = -\frac{1}{R_2C_2}\int -V_3 dt = \frac{V_3}{R_2C_2}t \tag{7.2}$$

DOI: 10.1201/9781003362968-7

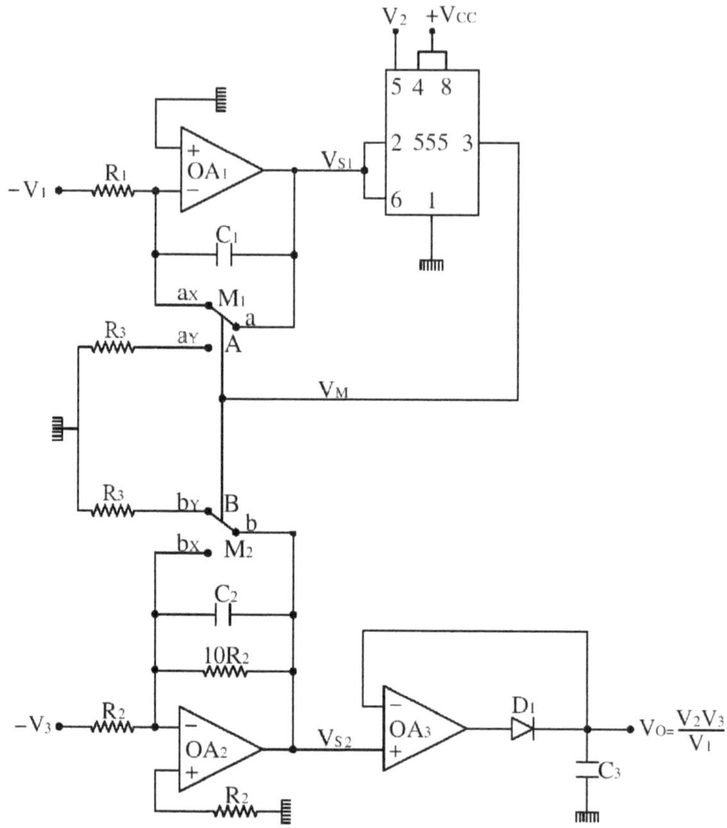

(a)

Figure 7.1 (a) Double single slope peak detecting MCD. (b) Double single slope peak sampling MCD.

Another saw tooth waveform V_{S2} with a peak value V_P is generated at the output of the op-amp OA_2. From the waveforms shown in Figure 7.2 and from equations (7.1) and (7.2), at t = T,

$$V_{S1} = V_2, V_{S2} = V_P$$

$$V_2 = \frac{V_1}{R_1C_1}T \quad (7.3)$$

$$V_P = \frac{V_3}{R_2C_2}T \quad (7.4)$$

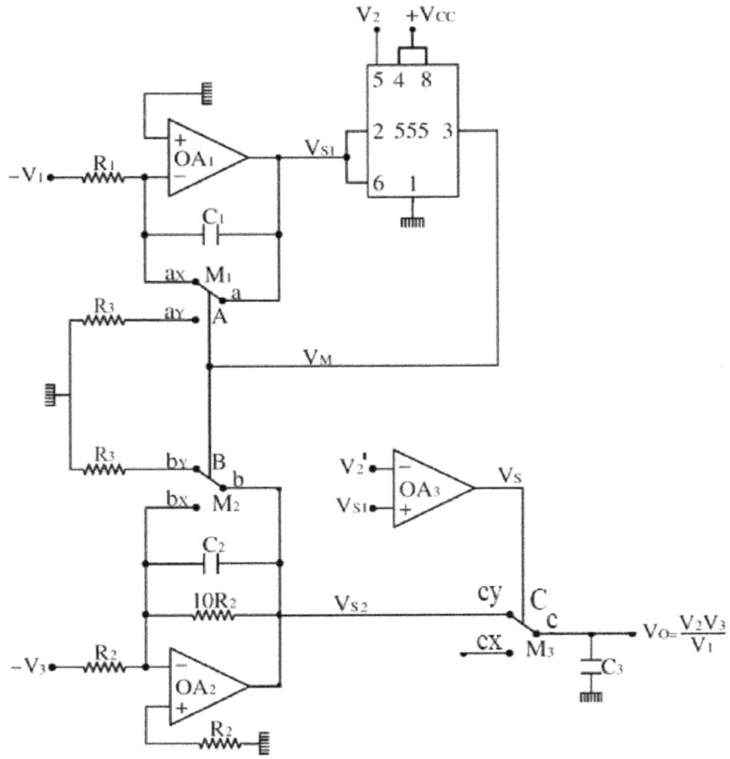

(b)

Figure 7.1 (Continued)

From equations (7.3) and (7.4),

$$V_P = \frac{V_3}{R_2C_2}\frac{V_2}{V_1}R_1C$$

Let us assume $R_1 = R_2$, $C_1 = C_2$. Then

$$V_P = \frac{V_2V_3}{V_1} \tag{7.5}$$

- In Figure 7.1(a), the peak detector realized by the op-amp OA_3, diode D_1, and capacitor C_3 gives this peak value V_P at its output V_O. $V_O = V_P$.

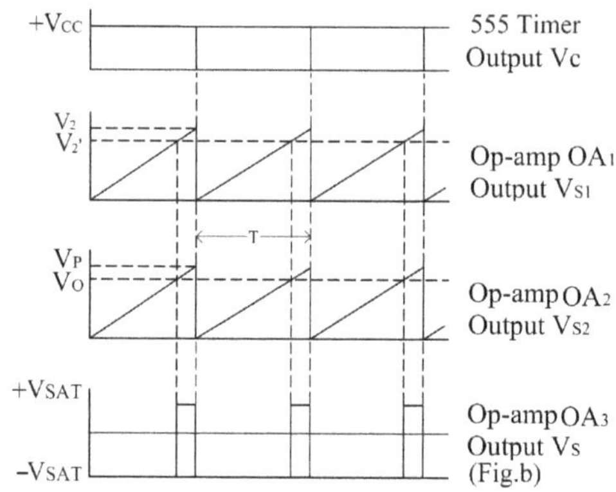

Figure 7.2 Associated waveforms of Figure 7.1.

- In Figure 7.1(b), the peak value V_P of the saw tooth waveform V_{S2} is obtained by the sample and hold circuit realized by the multiplexer M_3 and capacitor C_3. The sampling pulse is generated by the op-amp OA_3 by comparing a slightly lower voltage than that of V_2, called V_2', with the saw tooth wave V_{S1}. The sample and hold operation is illustrated graphically in Figure 7.2. The sample and hold output V_O is equal to V_P.

Hence the output will be $V_O = V_P$.

$$V_O = \frac{V_2 V_3}{V_1} \quad (7.6)$$

7.1.2 Double Single Slope—Type II

The circuit diagrams of double single slope peak responding MCDs are shown in Figure 7.3, and their associated waveforms are shown in Figure 7.4. Figure 7.3(a) shows double single slope peak detecting MCD, and Figure 7.3(b) shows double single slope peak sampling MCD. The 555 timer is configured to be a flip flop. Initially the 555 timer output is HIGH, the multiplexer M_1 connects ‘ay’ to ‘a’, an integrator, formed by the resistor R_1, capacitor C_1, and op-amp OA_1, integrates the first input voltage $-V_1$. The integrated output will be

$$V_{S1} = -\frac{1}{R_1 C_1}\int -V_1 dt = \frac{V_1}{R_1 C_1} t \quad (7.7)$$

A positive going ramp Vs_1 is generated at the output of op-amp OA_1. When the output of OA_1 reaches the voltage level of V_2, the comparator OA_2 output

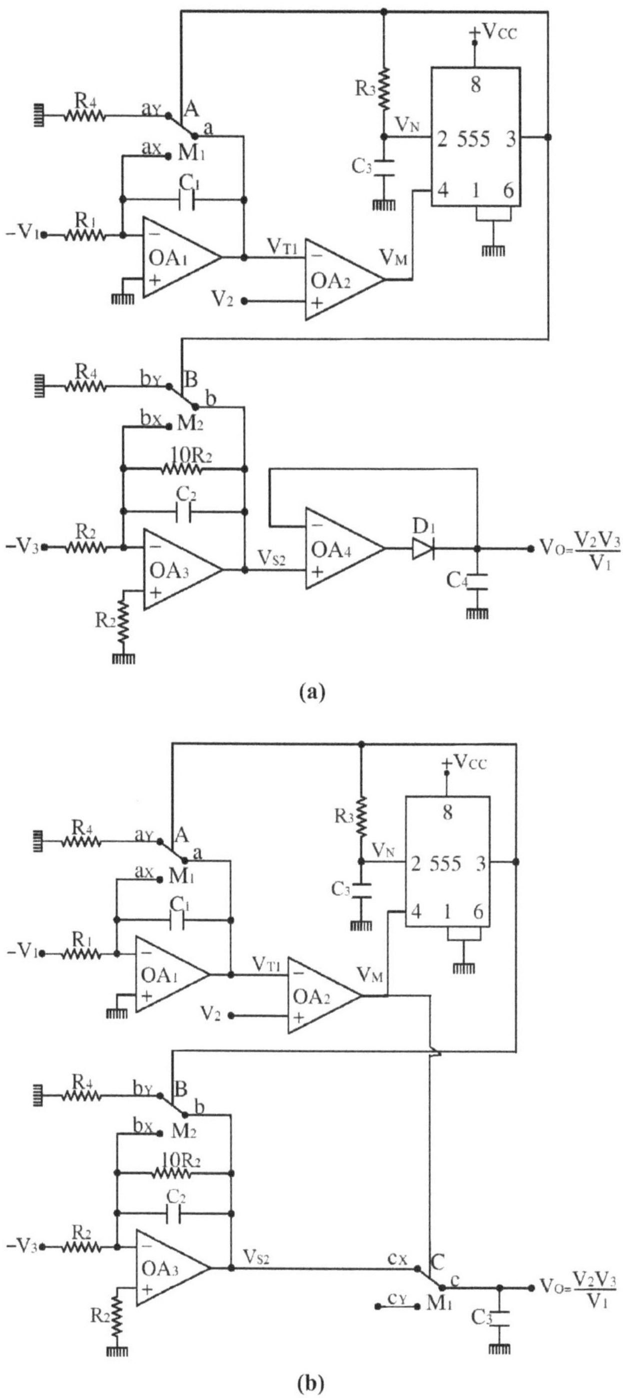

Figure 7.3 (a) Double single slope peak detecting MCD. (b) Double single slope peak sampling MCD.

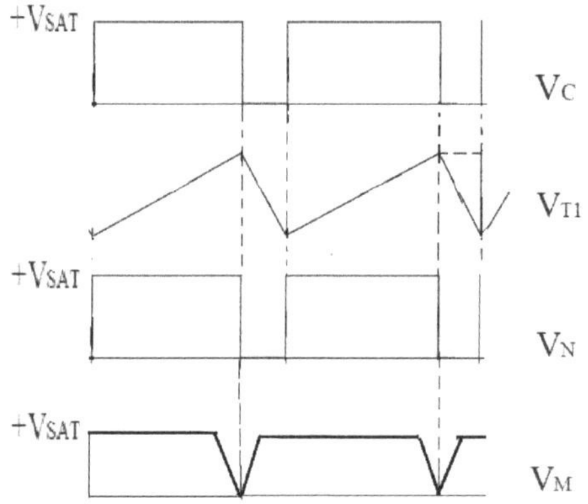

Figure 7.4 Associated waveforms of Figure 7.3.

becomes LOW and resets the 555 flip flop; the 555 timer output becomes LOW. The multiplexer M_1 connects 'ax' to 'a', and hence the capacitor C_1 is shorted so that the op-amp OA_1 output becomes zero. The 555 timer is set to HIGH from its output through the resistor R_3 and capacitor C_3, the multiplexer M_1 connects 'ay' to 'a', and the integrator composed of R_1, C_1, and op-amp OA_1 integrates the input voltage $-V_1$. The cycle therefore repeats to provide (1) a saw tooth wave of peak value V_2 at the output of op-amp OA_1 and (2) a short pulse waveform V_C at the output of the 555 timer. The short pulse V_C also controls the multiplexer M_2. During the short LOW time of V_C, the multiplexer M_2 connects 'bx' to 'b', the capacitor C_2 is short circuited so that the op-amp OA_3 output is zero volts. During the HIGH time of V_C, the multiplexer M_2 connects 'by' to 'b', the integrator, formed by the resistor R_2, capacitor C_2, op-amp OA_3, integrates its input voltage $-V_3$, and its output is given as

$$V_{S2} = -\frac{1}{R_2C_2}\int -V_3 dt = \frac{V_3}{R_2C_2}t \tag{7.8}$$

Another saw tooth waveform V_{S2} with a peak value V_P is generated at the output of the op-amp OA_3. From the waveforms shown in Figure 7.4 and from equations (7.7) and (7.8), at $t = T$, $V_{S1} = V_2$, $V_{S2} = V_P$.

$$V_2 = \frac{V_1}{R_1C_1}T \tag{7.9}$$

$$V_P = \frac{V_3}{R_2C_2}T \tag{7.10}$$

From equations (7.9) and (7.10),

$$V_P = \frac{V_3}{R_2C_2}\frac{V_2}{V_1}R_1C_1$$

Let us assume $R_1 = R_2$, $C_1 = C_2$. Then

$$V_P = \frac{V_2V_3}{V_1} \tag{7.11}$$

- In Figure 7.3(a), the peak detector realized by the op-amp OA_4, diode D_1, and capacitor C_3 gives this peak value V_P at its output V_O. $V_O = V_P$.
- In Figure 7.3(b), the peak value V_P of the saw tooth waveform V_{S2} is obtained by the sample and hold circuit realized by the multiplexer M_3 and capacitor C_3. The sample and hold operation is illustrated graphically in Figure 7.4. The sample and hold output V_O is equal to V_P.

Hence the output will be $V_O = V_P$:

$$V_O = \frac{V_2V_3}{V_1} \tag{7.12}$$

7.2 DOUBLE DUAL SLOPE PEAK RESPONDING MCD WITH FLIP FLOP

The circuit diagrams of double dual slope peak responding MCDs are shown in Figure 7.5, and their associated waveforms are shown in Figure 7.6. Figure 7.5(a) shows a peak detecting MCD, and Figure 7.5(b) shows a peak sampling MCD. Initially the flip flop output is LOW. The multiplexer M_1 connects $-V_1$ to the integrator I composed of the resistor R_1, capacitor C_1, and op-amp OA_1 ('ax' is connected to 'a'). The integrator I output is given as

$$V_{T1} = -\frac{1}{R_1C_1}\int(-V_1)dt = \frac{V_1}{R_1C_1}t \tag{7.13}$$

The output of integrator I is going toward positive saturation, and when it reaches the value $+V_2$, the comparator OA_2 output becomes LOW, and it sets the 555 flip flop output to HIGH. The multiplexer M_1 connects $+V_1$ to the integrator I composed of the resistor R_1, capacitor C_1, and op-amp OA_1 ('ay' is connected to 'a'). The integrator I output is given as

$$V_{T1} = -\frac{1}{R_1C_1}\int(+V_1)dt = -\frac{V_1}{R_1C_1}t \tag{7.14}$$

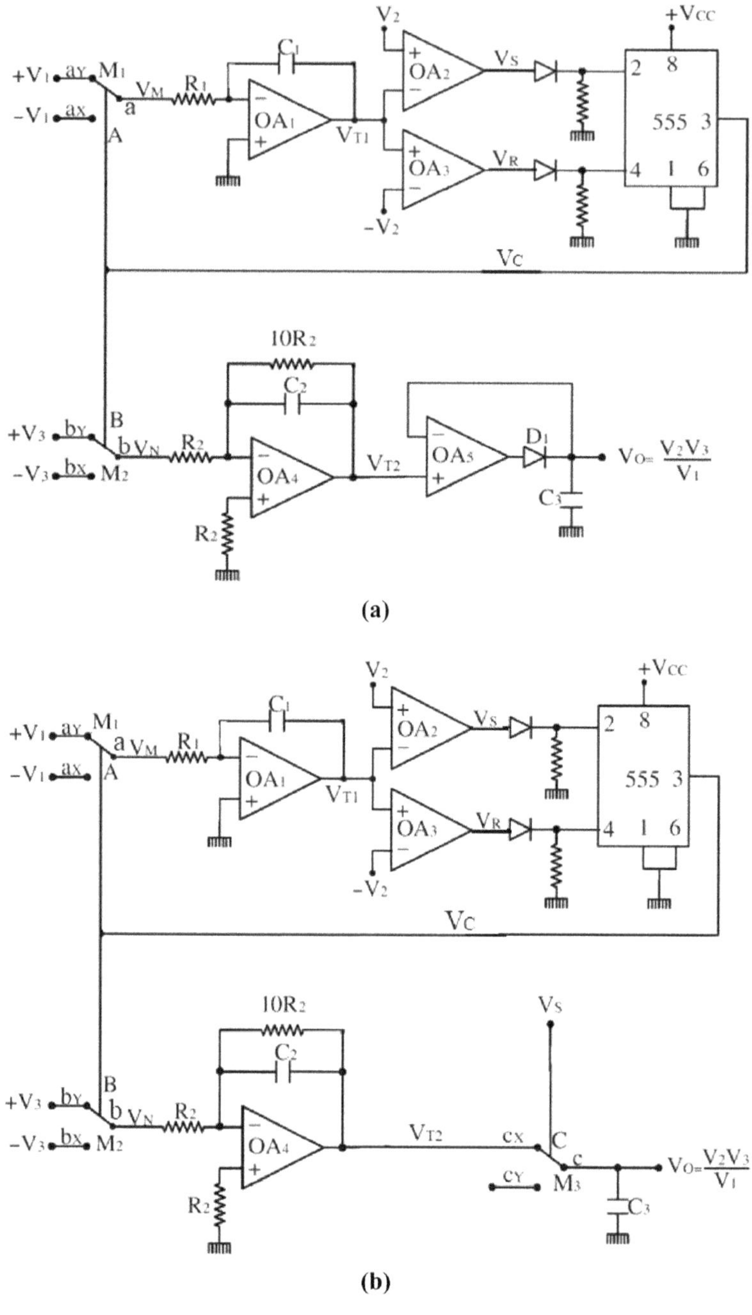

Figure 7.5 (a) Double dual slope peak detecting MCD with flip flop. (b) Double dual slope peak sampling MCD with flip flop.

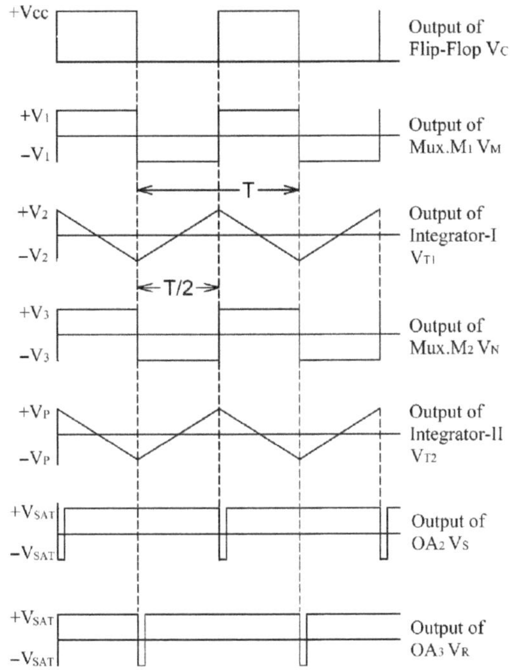

Figure 7.6 Associated waveforms of Figure 7.5.

The output of integrator I is reversing toward negative saturation, and when it reaches the value $-V_2$, the comparator OA_3 output becomes LOW, and it resets the 555 flip flop so that its output becomes LOW. The multiplexer M_1 connects $-V_1$, and the sequence repeats to give (1) a triangular waveform V_{T1} of $\pm V_2$ peak to peak values with a time period of T at the output of the op-amp OA_1, (2) a square waveform V_C at the output of the 555 flip flop, and (3) another square waveform V_M at the output of multiplexer M_1. From the waveforms shown in Figure 7.6, equation 7.13, and the fact that at $t = T/2$, $V_{T1} = 2V_2$.

$$2V_2 = \frac{V_1}{R_1C_1}\frac{T}{2}$$

$$T = \frac{4V_2}{V_1}R_1C_1 \tag{7.15}$$

The multiplexer M_2 connects $+V_3$ during the HIGH of the square waveform V_C ('by' is connected to 'b') and $-V_3$ during the LOW of the square waveform ('bx' is connected to 'b') V_C. Another square waveform V_N with

$\pm V_3$ peak to peak value is generated at the output of the multiplexer M_2. This square wave V_N is converted into a triangular wave V_{T2} by the integrator II composed of the resistor R_2, capacitor C_2, and op-amp OA_4 with $\pm V_P$ as the peak to peak values of the same time period T. For one transition, the integrator II output is given as

$$V_{T2} = -\frac{1}{R_2C_2}\int(-V_3)dt = \frac{V_3}{R_2C_2}t \tag{7.16}$$

From the waveforms shown in Figure 7.6, the equation (7.16), and the fact that at $t = T/2$, $V_{T2} = 2V_p$,

$$2V_P = \frac{V_3}{R_2C_2}\frac{T}{2}$$

$$V_P = \frac{V_2V_3}{V_1}\frac{R_1C_1}{R_2C_2}$$

Let $R_1 = R_2$ and $C_1 = C_2$:

$$V_P = \frac{V_2V_3}{V_1} \tag{7.17}$$

- In Figure 7.5(a), the peak detector realized by the op-amp OA_5, diode D_1, and capacitor C_3 gives the peak value V_p of triangular wave V_{T2}, and hence $V_O = V_P$.
- In Figure 7.5(b), the sample and hold circuit realized by the multiplexer M_3 and capacitor C_3 gives the peak value V_P of the triangular wave V_{T2}. The short pulse V_S generated at the input of the 555 timer acts as a sampling pulse. The sample and hold output is $V_O = V_P$.

From equation (7.17), $V_O = V_P$:

$$V_O = \frac{V_2V_3}{V_1} \tag{7.18}$$

7.3 PULSE WIDTH INTEGRATED PEAK RESPONDING MCD

The circuit diagrams of pulse width integrated peak responding MCDs are shown in Figure 7.7, and their associated waveforms are shown in Figure 7.8. Figure 7.7(a) shows a pulse width integrated peak detecting MCD, and Figure 7.7(b) shows a pulse width integrated peak sampling MCD. Initially the 555 timer output is HIGH, the multiplexer M_1 connects 'ay' to 'a',

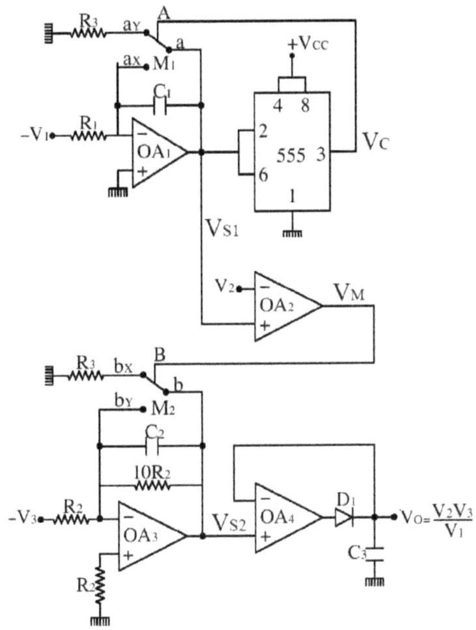

(a)

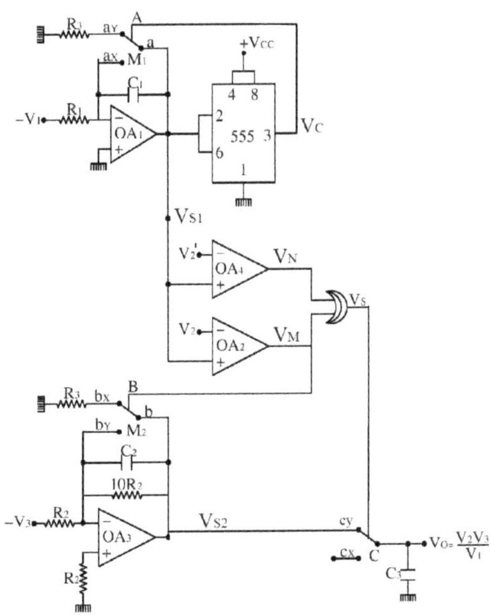

(b)

Figure 7.7 (a) Pulse width integrated peak detecting MCD. (b) Pulse width integrated peak sampled MCD.

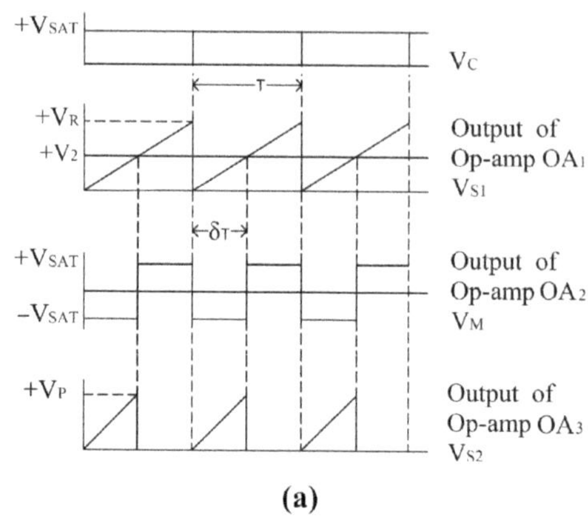

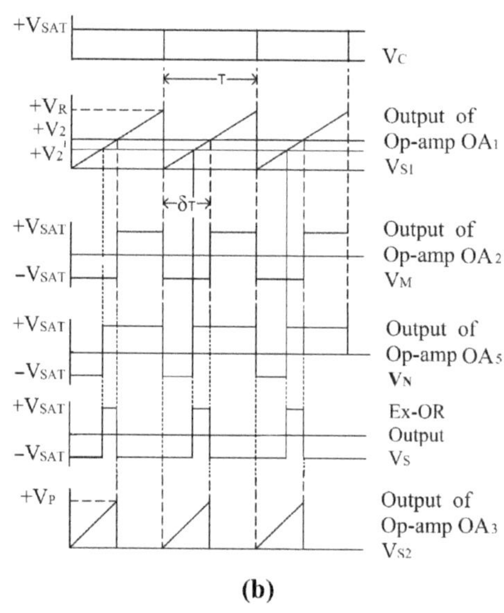

Figure 7.8 (a) Associated waveforms of Figure 7.7(a). (b) Associated waveforms of Figure 7.7(b).

and integrator, formed by the resistor R_1, capacitor C_1, and op-amp OA_1 integrates the first input voltage $-V_1$. The integrated output is given as

$$V_{S1} = -\frac{1}{R_1C_1}\int -V_1 dt = \frac{V_1}{R_1C_1}t \qquad (7.19)$$

When the output of op-amp OA_1 is rising toward positive saturation and reaches the value 2/3 V_{CC}, the 555 timer output becomes LOW, the multiplexer M_1 connects 'ax' to 'a', the capacitor C_1 is short circuited, and the op-amp OA_1 output becomes zero. Now the 555 timer output changes to HIGH, and the cycle therefore repeats to give (1) a saw tooth waveform V_{S1} of peak value V_R and time period T at the output of op-amp OA_1 and (2) a short pulse wave form V_C at the output of the 555 timer. From the waveforms shown in Figure 7.8 and the fact that at $t = T$, $V_{S1} = V_R = 2/3\ V_{CC}$.

$$V_R = \frac{V_1}{R_1C_1}T, T = \frac{V_R}{V_1}R_1C_1 \tag{7.20}$$

The saw tooth waveform V_{S1} is compared with the second input voltage V_2 by the comparator OA_2. An asymmetrical rectangular wave V_M is generated at the output of comparator OA_2. The OFF time of this wave V_M is given as

$$\delta_T = \frac{V_2}{V_R}T \tag{7.21}$$

The output of the comparator OA_2 is given as control input of the multiplexer M_2. During the ON time of V_M, the multiplexer M_2 connects 'by' to 'b', the capacitor C_2 is shorted so that zero voltage appears at op-amp OA_3 output. During the OFF time of V_M, the multiplexer M_2 connects 'bx' to 'b', another integrator is formed by resistor R_2, capacitor C_2, and op-amp OA_3. This integrator integrates the third input voltage $-V_3$, and its output is given as

$$V_{S2} = -\frac{1}{R_2C_2}\int -V_3 dt = \frac{V_3}{R_2C_2}t \tag{7.22}$$

A semi-saw tooth wave V_{S2} with peak values of V_P is generated at the output of the op-amp OA_3. From the waveforms shown in Figure 7.8, equation (7.22), and fact that at $t = \delta_T$, $V_{S2} = V_P$.

$$V_P = \frac{V_3}{R_2C_2}\delta_T \tag{7.23}$$

$$V_P = \frac{V_2V_3}{V_1}\frac{R_1C_1}{R_2C_2} \tag{7.24}$$

Let us assume $R_1C_1 = R_2C_2$

$$V_P = \frac{V_2V_3}{V_1} \tag{7.25}$$

- In Figure 7.7(a), peak detector realized by the diode D_1 and capacitor C_3 gives the peak value V_P at its output. $V_0 = V_P$.
- In Figure 7.7(b), the peak value V_P is obtained by the sample and hold circuit realized by the multiplexer M_3 and capacitor C_3. The sampling pulse V_S is generated by the Ex-OR gate from the signals V_M and V_N. V_N is obtained by comparing a slightly lower voltage than that of V_2, i.e., V_2', with the saw tooth waveform V_{S1}. The sampled output is given as $V_O = V_P$.

$$V_O = \frac{V_2 V_3}{V_1} \tag{7.26}$$

7.4 PULSE POSITION PEAK RESPONDING MCDs

The circuit diagrams of pulse position peak responding MCDs are shown in Figure 7.9, and their associated waveforms are shown in Figure 7.10. Figure 7.9(a) shows a pulse position peak detecting MCD, and Figure 7.9(b) shows a pulse position peak sampling MCD.

The op-amp OA_1 and 555 timer constitute a saw tooth wave generator. The time period of the saw tooth wave V_{S1} is given as

$$T = \frac{V_R}{V_1} R_1 C_1 \tag{7.27}$$

The comparator OA_3 compares the saw tooth wave V_{S1} with an input voltage V_2 and produces a rectangular wave V_M. The OFF time δ_T of this rectangular wave V_M is given as

$$\delta_T = \frac{V_2}{V_R} T \tag{7.28}$$

where $V_R = 2V_{CC}/3$.

The short pulse V_C in the saw tooth wave generator is also given to the multiplexer M_2, which constitutes a controlled integrator along with the op-amp OA_3, resistor R_2, and capacitor C_2. During the HIGH value of V_C, the multiplexer M_2 connects 'by' to 'b', another integrator is formed by the op-amp OA_2, resistor R_2, and capacitor C_2. During the LOW value of V_C, the multiplexer M_2 connects 'bx' to 'b', and capacitor C_2 is short circuited so that integrator OA_2 output becomes zero.

The integrator OA_2 output is given as

$$V_{S2} = -\frac{1}{R_2 C_2} \int -V_3 dt = \frac{V_3}{R_2 C_2} t \tag{7.29}$$

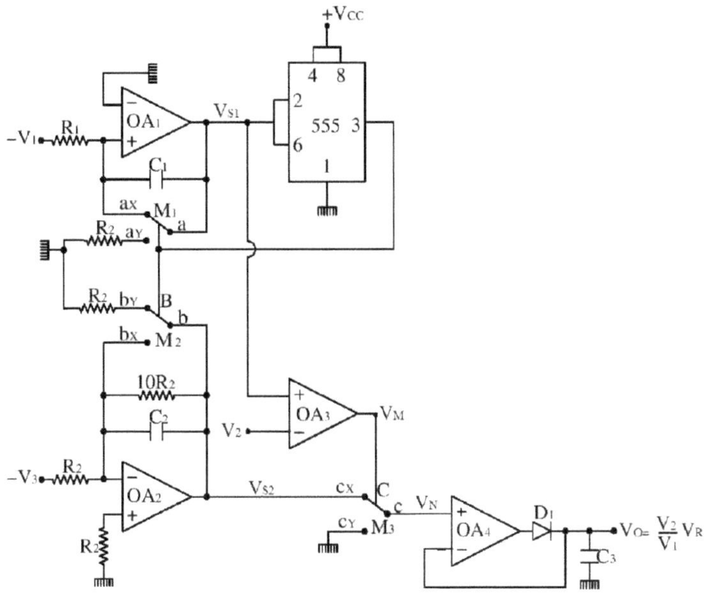

(a)

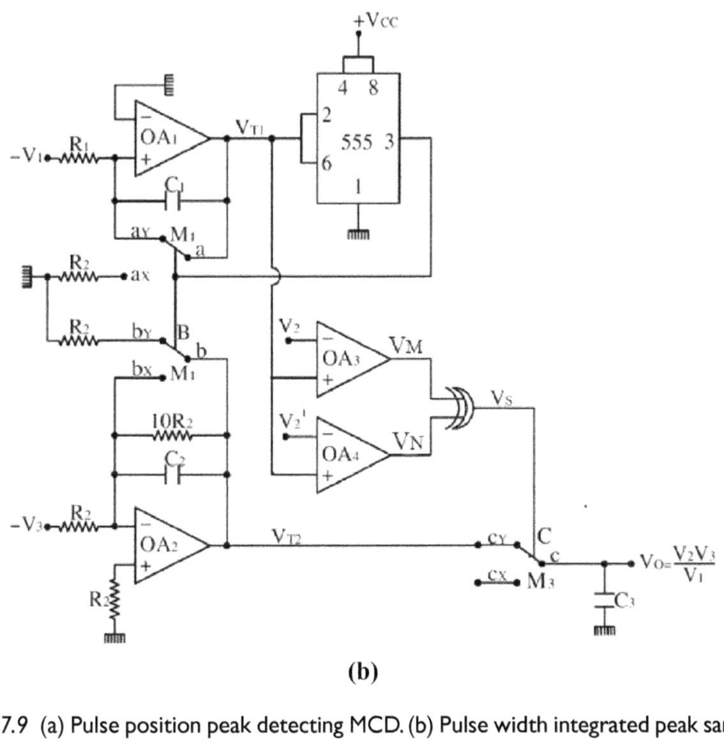

(b)

Figure 7.9 (a) Pulse position peak detecting MCD. (b) Pulse width integrated peak sampling MCD.

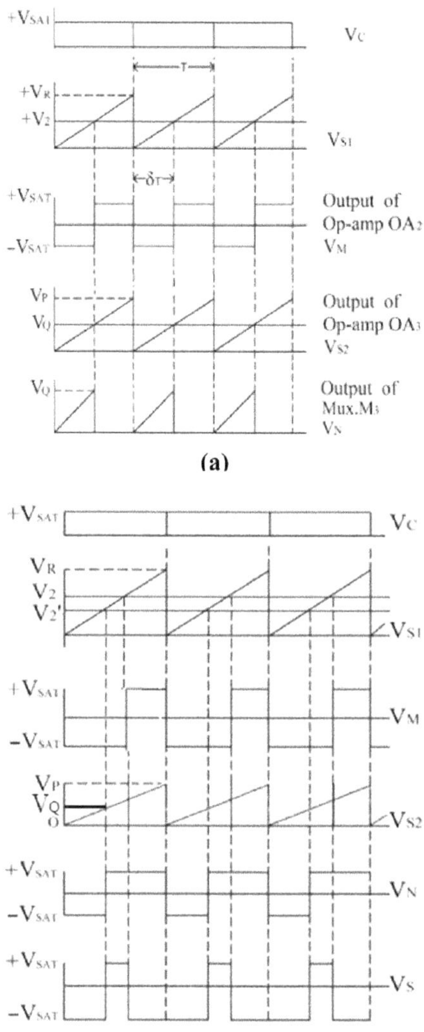

Figure 7.10 (a) Associated waveforms of Figure 7.9(a). (b) Associated waveforms of Figure 7.9(b).

Another saw tooth wave V_{S2} with a peak value of V_P is generated at the output of integrator OA_3. From the waveforms shown in Figure 7.10, equation (7.29), and the fact that at t = T, $V_{S2} = V_P$:

$$V_P = \frac{V_3}{R_2C_2}T \tag{7.30}$$

- In Figure 7.9(a), the rectangular pulse V_M controls the multiplexer M_3. During the LOW of V_M, the multiplexer M_3 connects 'cx' to 'c', and the saw tooth wave V_{S2} is connected to the multiplexer M_3 output. During the HIGH of V_M, the multiplexer M_3 connects 'cy' to 'c', and zero voltage is connected to the multiplexer M_3. A semi-saw tooth wave V_N with peak value V_Q is generated at the output of the multiplexer M_3. The peak detector realized by the op-amp OA_4, diode D_1, and capacitor C_3 gives the peak value V_Q at its output, i.e., $V_O = V_Q$.
- In Figure 7.9(b), the saw tooth wave V_{S2} is sampled by the sample and hold circuit realized by the multiplexer M_3 and capacitor C_3 with a sampling pulse V_S. The sampling pulse V_S is generated by Ex-ORing the V_M and V_N signals. The signal V_M is generated by comparing the first saw tooth wave V_{S1} with the second input voltage V_2. The signal V_N is generated by comparing the first saw tooth wave V_{S1} with a slightly lower voltage than that of V_2, i.e., V_2'. The sample and hold output $V_O = V_Q$.

The value of V_Q is given as

$$V_Q = \frac{V_P}{T}\delta_T \tag{7.31}$$

$$V_O = \frac{V_3}{R_2C_2}\frac{V_2}{V_R}T$$

$$V_O = \frac{V_2V_3}{V_1}\frac{R_1C_1}{R_2C_2}$$

Let $R_1 = R_2$, $C_1 = C_2$. Then

$$V_O = \frac{V_2V_3}{V_1} \tag{7.32}$$

Chapter 8

Peak Responding Multiplier cum Dividers—Switching

8.1 DOUBLE SINGLE SLOPE PEAK RESPONDING MCD

8.1.1 Type I

The circuit diagrams of double single slope peak responding MCDs are shown in Figure 8.1, and their associated waveforms are shown in Figure 8.2. Figure 8.1(a) shows a double single slope peak detecting MCD, and Figure 8.1(b) shows a double single slope peak sampling MCD. Initially the 555 timer output is HIGH, the switch S_1 is opened, and an integrator, formed by the resistor R_1, capacitor C_1, and op-amp OA_1 integrates the first input voltage $-V_1$. The integrated output will be

$$V_{S1} = -\frac{1}{R_1C_1}\int -V_1 dt = \frac{V_1}{R_1C_1}t \tag{8.1}$$

A positive going ramp Vs_1 is generated at the output of the op-amp OA_1. When the output of OA_1 reaches the voltage level of V_2, the 555 timer output becomes LOW. The switch S_1 is closed, and hence the capacitor C_1 is shorted so that the op-amp OA_1 output becomes zero. Then the 555 timer output goes to HIGH, the switch S_1 is opened, and the integrator, composed of the R_1, C_1, and op-amp OA_1, integrates the input voltage $-V_1$, and the cycle therefore repeats to provide (1) a saw tooth wave V_{S1} of peak value V_2 at the output of the op-amp OA_1 and (2) a short pulse waveform V_C at the output of the 555 timer. The short pulse V_C also controls the switch S_2. During the short LOW time of V_C, the switch S_2 is closed, and the capacitor C_2 is short circuited so that op-amp OA_2 output is zero volts. During the HIGH time of V_C, the switch S_2 is opened, and the integrator, formed by the resistor R_2, capacitor C_2, op-amp OA_2, integrates its input voltage $-V_3$. Its output is given as

$$V_{S2} = -\frac{1}{R_2C_2}\int -V_3 dt = \frac{V_3}{R_2C_2}t \tag{8.2}$$

DOI: 10.1201/9781003362968-8

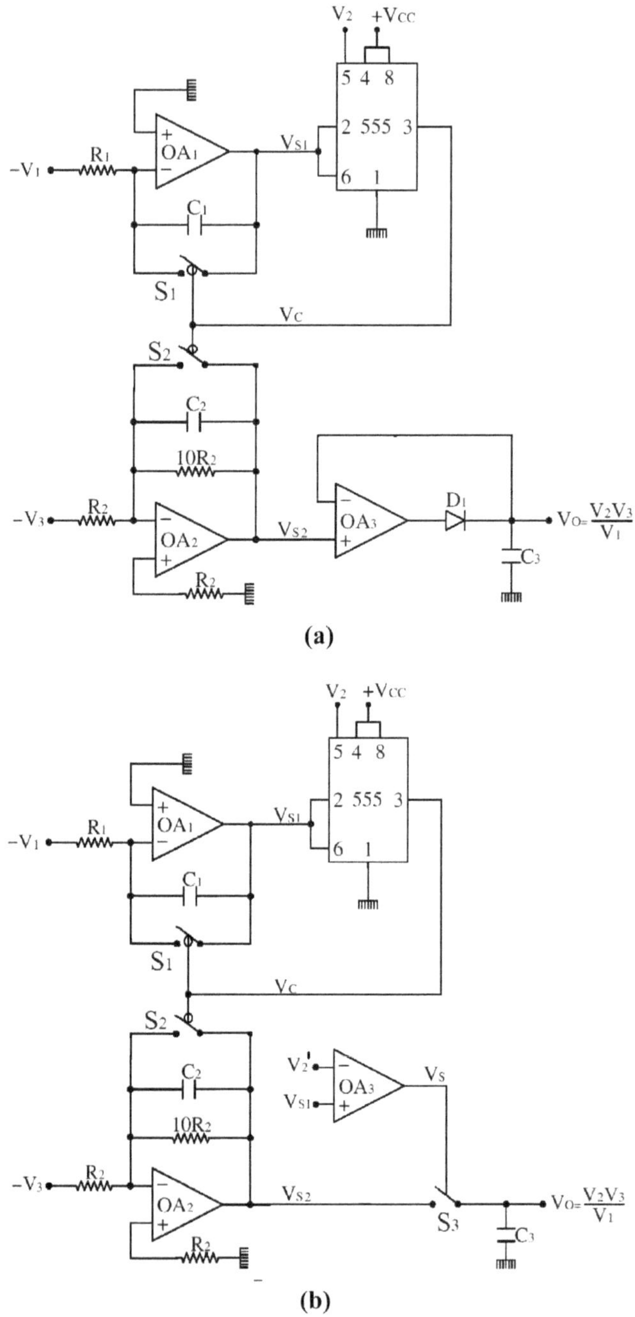

Figure 8.1 (a) Double single slope peak detecting MCD. (b) Double single slope peak sampling MCD.

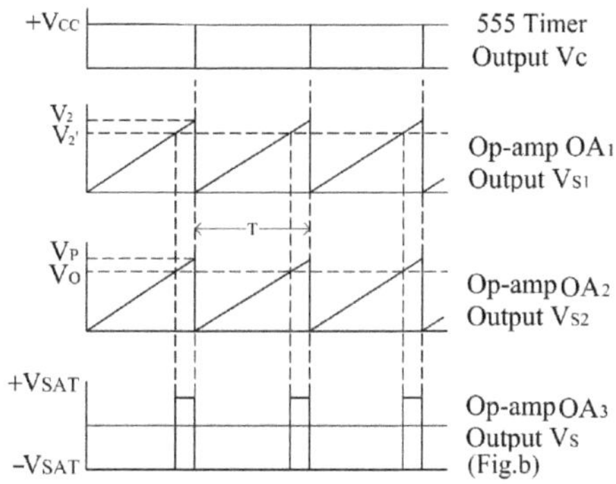

Figure 8.2 Associated waveforms of Figure 8.1.

Another saw tooth waveform V_{S2} with peak value V_P is generated at the output of the op-amp OA_3. From the waveforms shown in Figure 8.2 and equations (8.1) and (8.2), at t = T, $V_{S1} = V_2$, $V_{S2} = V_P$.

$$V_2 = \frac{V_1}{R_1C_1}T \tag{8.3}$$

$$V_P = \frac{V_3}{R_2C_2}T \tag{8.4}$$

From equations (8.3) and (8.4),

$$V_P = \frac{V_3}{R_2C_2}\frac{V_2}{V_1}R_1C_1$$

Let us assume $R_1 = R_2$, $C_1 = C_2$. Then

$$V_P = \frac{V_2V_3}{V_1} \tag{8.5}$$

- In Figure 8.1(a), the peak detector realized by the op-amp OA_3, diode D_1, and capacitor C_3 gives this peak value V_P at its output V_O. $V_O = V_P$.
- In Figure 8.1(b), the peak value V_P of the saw tooth waveform V_{S2} is obtained by the sample and hold circuit realized by the switch S_3 and capacitor C_3. The sampling pulse is generated by the op-amp OA_3 by comparing a slightly lower voltage than that of V_2, called V_2', with the saw tooth wave V_{S1}. The sample and hold operation is illustrated graphically in Figure 8.2. The sample and hold output V_O is equal to V_P.

Hence the output will be $V_O = V_P$:

$$V_O = \frac{V_2 V_3}{V_1} \tag{8.6}$$

8.1.2 Double Single Slope—Type II

The circuit diagrams of double single slope peak responding MCDs are shown in Figure 8.3, and their associated waveforms are shown in Figure 8.4. Figure 8.3(a) shows a double single slope peak detecting MCD, and Figure 8.3(b) shows a double single slope peak sampling MCD. The 555 timer is configured to be a flip flop. Initially the 555 timer output is HIGH, the switch S_1 is opened, and an integrator, formed by the resistor R_1, capacitor C_1, and op-amp OA_1, integrates the first input voltage $-V_1$. The integrated output will be

$$V_{S1} = -\frac{1}{R_1 C_1}\int -V_1 dt = \frac{V_1}{R_1 C_1} t \tag{8.7}$$

A positive going ramp Vs_1 is generated at the output of the op-amp OA_1. When the output of OA_1 reaches the voltage level of V_2, the comparator

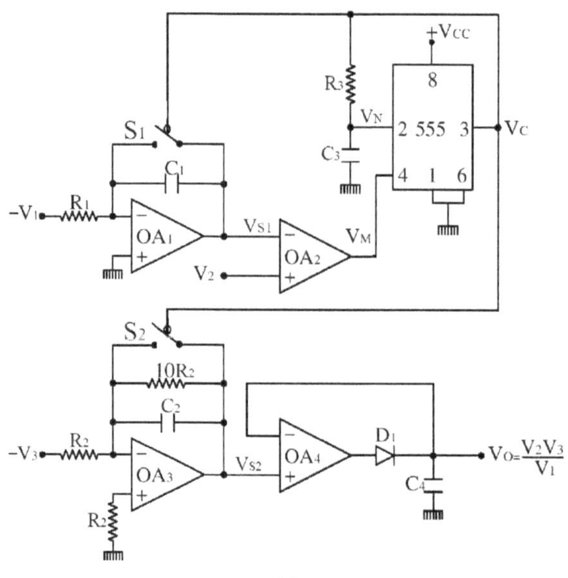

(a)

Figure 8.3 (a) Double single slope peak detecting MCD. (b) Double single slope peak sampling MCD.

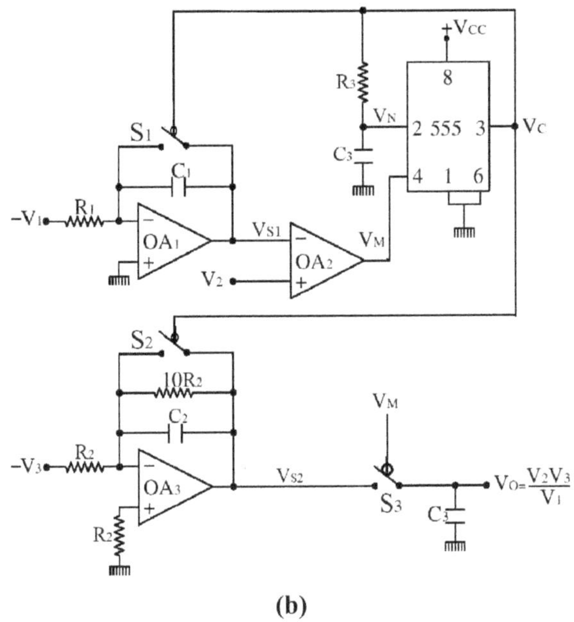

Figure 8.3 (Continued)

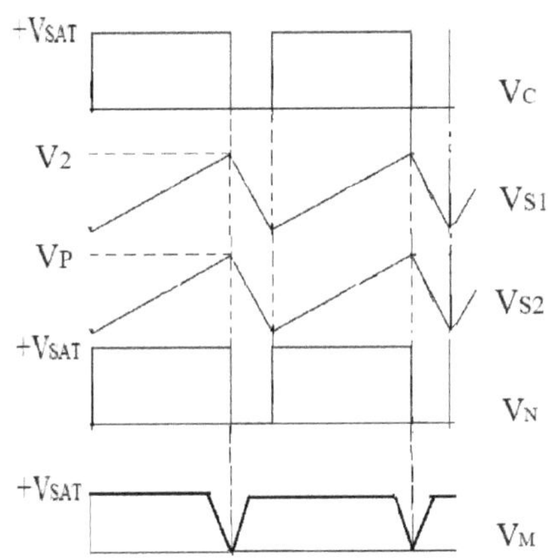

Figure 8.4 Associated waveforms of Figure 8.3.

OA_2 output becomes LOW and resets the 555 flip flop, and the 555 timer output becomes LOW. The switch S_1 is closed, and hence the capacitor C_1 is shorted so that the op-amp OA_1 output becomes zero. The 555 timer is set to HIGH from its output through resistor R_3 and capacitor C_3, the switch S_1 is opened, and the integrator, composed of R_1, C_1, and op-amp OA_1, integrates the input voltage $-V_1$. The cycle therefore repeats to provide (1) a saw tooth wave of peak value V_2 at the output of the op-amp OA_1 and (2) a short pulse waveform V_C at the output of the 555 timer. The short pulse V_C also controls the switch S_2. During the short LOW time of V_C, the switch S_2 is closed, and the capacitor C_2 is short circuited so that op-amp OA_3 output is zero voltage. During the HIGH time of V_C, switch S_2 is opened, and the integrator, formed by resistor R_2, capacitor C_2, op-amp OA_3, integrates its input voltage $-V_3$. Its output is given as

$$V_{S2} = -\frac{1}{R_2C_2}\int -V_3 dt = \frac{V_3}{R_2C_2}t \qquad (8.8)$$

Another saw tooth waveform V_{S2} with peak value V_P is generated at the output of the op-amp OA_3. From the waveforms shown in Figure 8.5 and from equations (8.7) and (8.8), at $t = T$, $V_{S1} = V_2$, $V_{S2} = V_P$.

$$V_2 = \frac{V_1}{R_1C_1}T \qquad (8.9)$$

$$V_P = \frac{V_3}{R_2C_2}T \qquad (8.10)$$

From equations (8.9) and (8.10),

$$V_P = \frac{V_3}{R_2C_2}\frac{V_2}{V_1}R_1C_1$$

Let us assume $R_1 = R_2$, $C_1 = C_2$. Then

$$V_P = \frac{V_2V_3}{V_1} \qquad (8.11)$$

- In Figure 8.3(a), the peak detector realized by the op-amp OA_4, diode D_1, and capacitor C_3 gives the peak value V_P at its output V_O. $V_O = V_P$.
- In Figure 8.3(b), the peak value V_P of the saw tooth waveform V_{S2} is obtained by the sample and hold circuit realized by the switch S_3 and capacitor C_3. The sample and hold operation is illustrated graphically in Figure 8.4. The sample and hold output V_O is equal to V_P.

Hence the output will be $V_O = V_P$.

$$V_O = \frac{V_2V_3}{V_1} \qquad (8.12)$$

8.2 DOUBLE DUAL SLOPE PEAK RESPONDING MCD WITH FLIP FLOP

The circuit diagram of double dual slope peak responding MCDs are shown in Figure 8.5, and their associated waveforms are shown in Figure 8.6. Figure 8.5(a) shows a peak detecting MCD, and Figure 8.5(b) shows a peak

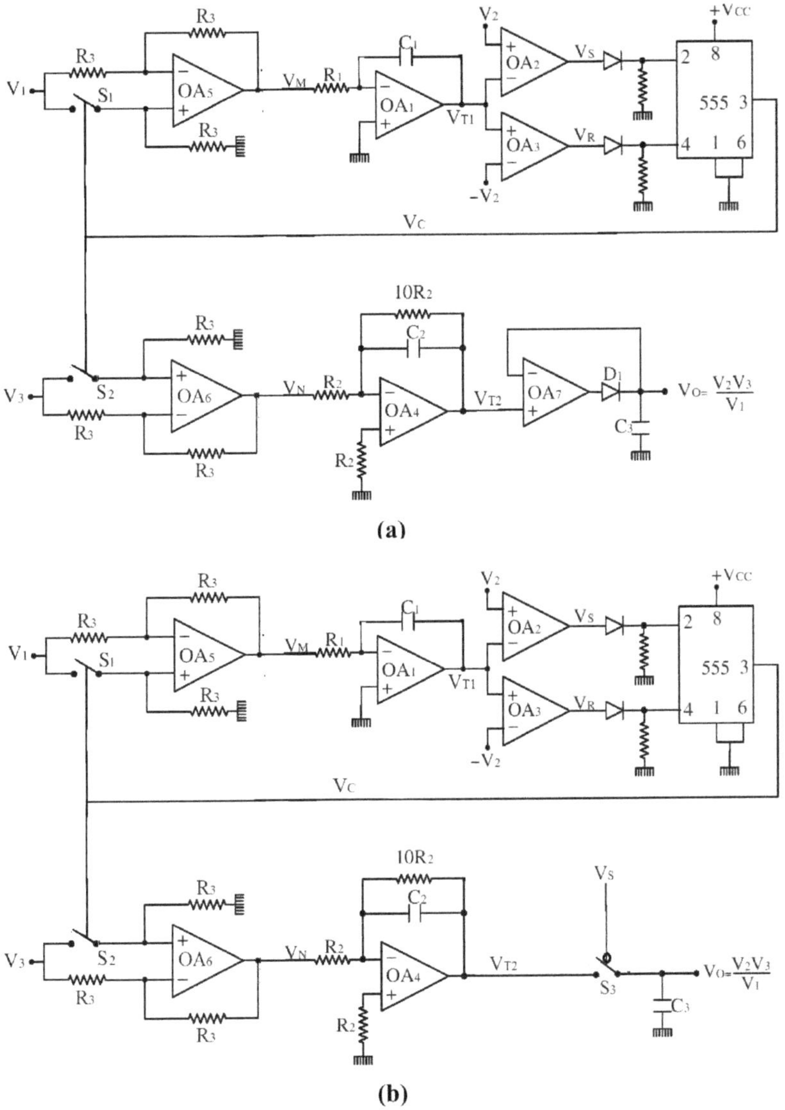

Figure 8.5 (a) Double dual slope peak detecting MCD with flip flop. (b) Double dual slope peak sampling MCD with flip flop.

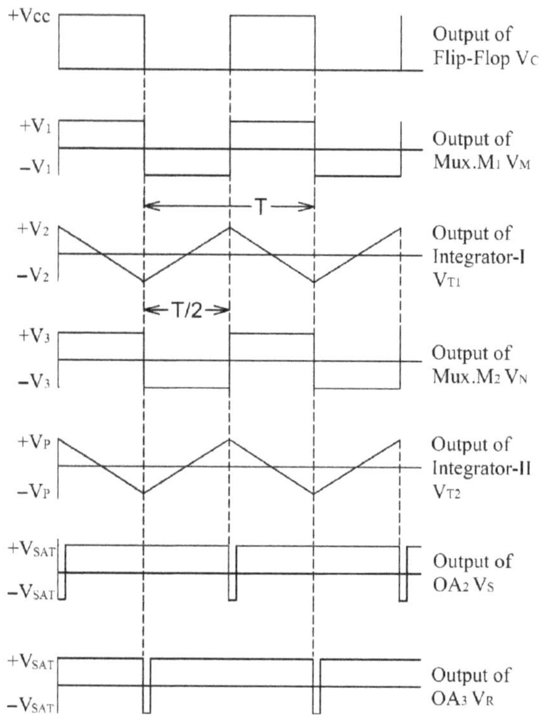

Figure 8.6 Associated waveforms of Figure 8.5.

sampling MCD. Initially the flip flop output is LOW. The control amplifier OA_5 gives $-V_1$ to the integrator I, composed of the resistor R_1, capacitor C_1, and op-amp OA_1 (the switch S_1 is opened, and the op-amp OA_5 will work as an inverting amplifier). The integrator I output is given as

$$V_{T1} = -\frac{1}{R_1C_1}\int(-V_1)dt = \frac{V_1}{R_1C_1}t \tag{8.13}$$

The output of integrator I is going toward positive saturation, and when it reaches the value $+V_2$, the comparator OA_2 output becomes LOW, and it sets the 555 flip flop output to HIGH. The control amplifier OA_5 gives $+V_1$ to the integrator I composed by resistor R_1, capacitor C_1, and op-amp OA_1 (the switch S_1 is closed, and the op-amp OA_5 will work as a non-inverting amplifier). The integrator I output is given as

$$V_{T1} = -\frac{1}{R_1C_1}\int(+V_1)dt = -\frac{V_1}{R_1C_1}t \tag{8.14}$$

The output of integrator I is reversing toward negative saturation, and when it reaches the value $-V_2$, the comparator OA_3 output becomes LOW and resets the 555 flip flop, so that its output becomes LOW. The control amplifier OA_5 connects $-V_1$, and the sequence repeats to give (1) a triangular waveform V_{T1} of $\pm V_2$ peak to peak values with a time period of T at the output of the op-amp OA_1, (2) a square waveform V_C at the output of the 555 flip flop, and (3) another square waveform V_M at the output of the control amplifier OA_5. From the waveforms shown in Figure 8.8, equation (8.13), and the fact that at $t = T/2$, $V_{T1} = 2V_2$,

$$2V_2 = \frac{V_1}{R_1C_1}\frac{T}{2}$$

$$T = \frac{4V_2}{V_1}R_1C_1 \tag{8.15}$$

The control amplifier OA_6 gives $+V_3$ during the HIGH of the square waveform V_C (the switch S_2 is closed, and the control amplifier OA_5 will work as a non-inverting amplifier) and $-V_3$ during the LOW of the square waveform (the switch S_2 is opened, and the control amplifier OA_6 will work as an inverting amplifier) V_C. Another square waveform V_N with $\pm V_3$ peak to peak value is generated at the output of the op-amp OA_6. This square wave V_N is converted to the triangular wave V_{T2} by the integrator II, composed of the resistor R_2, capacitor C_2, and op-amp OA_4 with $\pm V_P$ as the peak to peak values of the same time period T. For one transition, the integrator II output is given as

$$V_{T2} = -\frac{1}{R_2C_2}\int(-V_3)dt = \frac{V_3}{R_2C_2}t \tag{8.16}$$

From the waveforms shown in Figure 8.6, equation (8.16), and the fact that at $t = T/2$, $V_{T2} = 2V_p$.

$$2V_P = \frac{V_3}{R_2C_2}\frac{T}{2}$$

$$V_P = \frac{V_2V_3}{V_1}\frac{R_1C_1}{R_2C_2}$$

Let $R_1 = R_2$ and $C_1 = C_2$.

$$V_P = \frac{V_2V_3}{V_1} \tag{8.17}$$

- In Figure 8.5(a), the peak detector realized by the op-amp OA_7, diode D_1, and capacitor C_3 gives the peak value V_p of the triangular wave V_{T2}, and hence $V_O = V_P$.
- In Figure 8.5(b), the sample and hold circuit realized by the switch S_3 and capacitor C_3 gives the peak value V_P of the triangular wave V_{T2}. The short pulse V_S generated at the input of the 555 timer acts as a sampling pulse. The sample and hold output is $V_O = V_P$.

From equation (8.17), $V_O = V_P$.

$$V_O = \frac{V_2 V_3}{V_1} \tag{8.18}$$

8.3 PULSE WIDTH INTEGRATED PEAK RESPONDING MCD

The circuit diagrams of pulse width integrated peak responding MCDs are shown in Figure 8.7, and their associated waveforms are shown in Figure 8.8. Figure 8.7(a) shows a pulse width integrated peak detecting MCD, and Figure 8.7(b) shows a pulse width integrated peak sampling MCD. Initially the 555 timer output is HIGH, the switch S_1 is opened, and the integrator formed by the resistor R_1, capacitor C_1, and op-amp OA_1, integrates $-V_1$. The integrated output is given as

$$V_{S1} = -\frac{1}{R_1C_1}\int -V_1 dt = \frac{V_1}{R_1C_1}t \tag{8.19}$$

When the output of the op-amp OA_1 is rising toward positive saturation and reaches the value 2/3 V_{CC}, the 555 timer output will become LOW, the switch S_1 is closed, the capacitor C_1 is short circuited, and op-amp OA_1 output becomes zero. Now the 555 timer output changes to HIGH, and the cycle therefore repeats to give (1) a saw tooth waveform V_{S1} of peak value V_R and time period T at the output of the op-amp OA_1 and (2) a short pulse waveform V_C at the output of the 555 timer. From the waveforms shown in Figure 8.11 and the fact that at t = T, $V_{S1} = V_R = 2/3\ V_{CC}$,

$$V_R = \frac{V_1}{R_1C_1}T, T = \frac{V_R}{V_1}R_1C_1 \tag{8.20}$$

The saw tooth waveform V_{S1} is compared with the second input voltage V_2 by the comparator OA_2. An asymmetrical rectangular wave V_M is

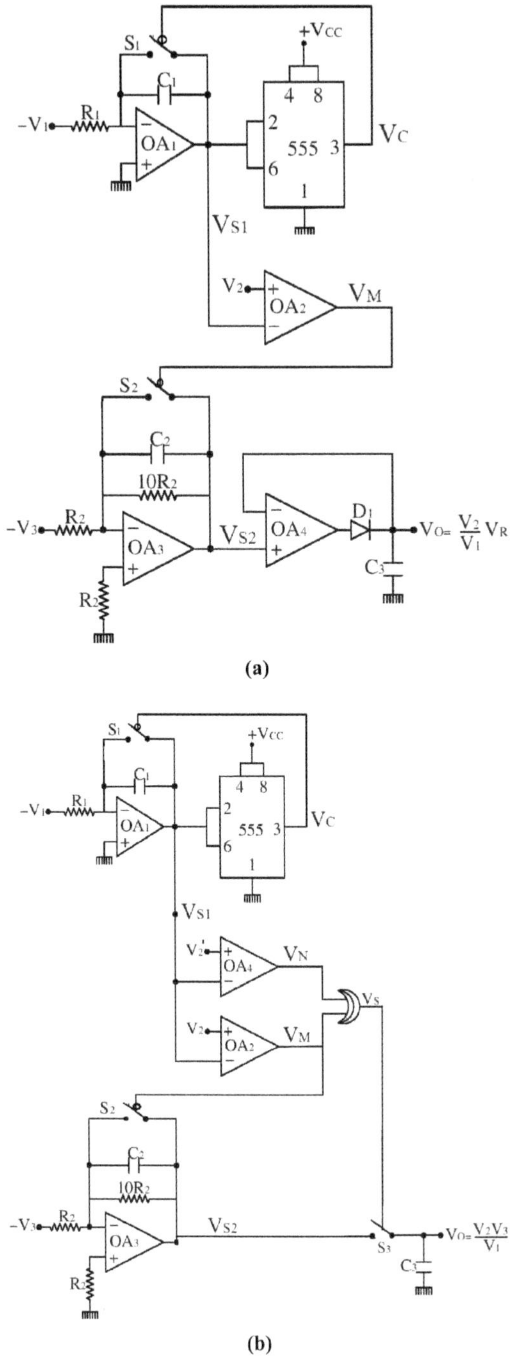

Figure 8.7 (a) Pulse width integrated peak detecting MCD. (b) P ulse width integrated peak sampling MCD.

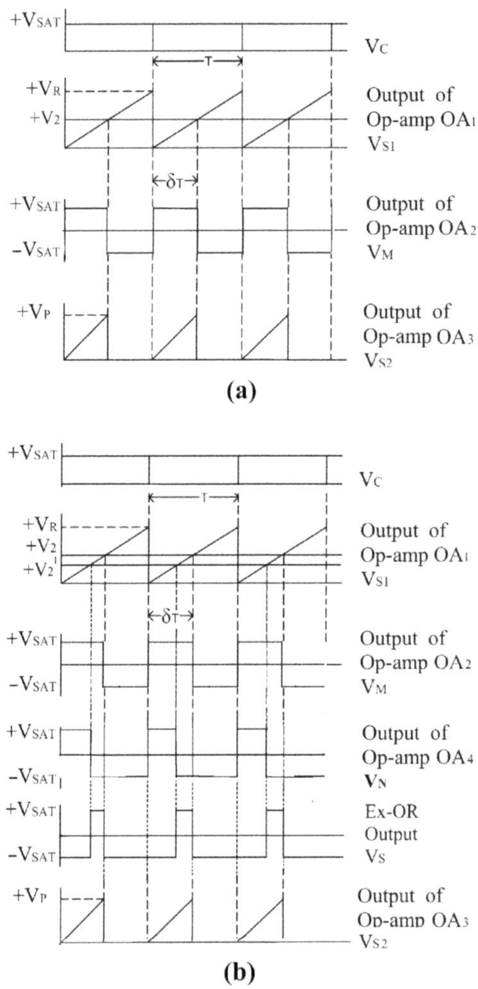

Figure 8.8 (a) Associated waveforms of Figure 8.7(a). (b) Associated waveforms of Figure 8.7(b).

generated at the output of the comparator OA_2. The ON time of this wave V_M is given as

$$\delta_T = \frac{V_2}{V_R} T \tag{8.21}$$

The output of the comparator OA_2 is given as control input of the switch S_2. During the OFF time of V_M, the switch S_2 is closed, the capacitor C_2 is shorted so that zero voltage appears at the op-amp OA_3 output. During the ON time of V_M, the switch S_2 is opened, another integrator is formed by resistor R_2, capacitor C_2, and op-amp OA_3. This integrator integrates the third input voltage $-V_3$, and its output is given as

$$V_{S2} = -\frac{1}{R_2C_2}\int -V_3 dt = \frac{V_3}{R_2C_2}t \tag{8.22}$$

A semi-saw tooth wave V_{S2} with peak values of V_P is generated at the output of op-amp OA_3. From the waveforms shown in Figure 8.8, equation (8.22), and fact that at $t = \delta_T$, $V_{S2} = V_P$.

$$V_P = \frac{V_3}{R_2C_2}\delta_T$$

$$V_P = \frac{V_2V_3}{V_1}\frac{R_1C_1}{R_2C_2}$$

Let us assume $R_1C_1 = R_2C_2$.

$$V_P = \frac{V_2V_3}{V_1} \tag{8.23}$$

- In Figure 8.7(a), the peak detector realized by the op-amp OA_4, diode D_1, and capacitor C_3 gives the peak value V_P at its output. $V_0 = V_P$.
- In Figure 8.7(b), the peak value V_P is obtained by the sample and hold circuit realized by the switch S_3 and capacitor C_3. The sampling pulse V_S is generated by the Ex-OR gate from the signals V_M and V_N. V_N is obtained by comparing slightly lower voltage than that of V_2, i.e., V_2', with the saw tooth waveform V_{S1}. The sampled output is given as $V_O = V_P$.

$$V_O = \frac{V_2V_3}{V_1} \tag{8.24}$$

8.4 PULSE POSITION PEAK RESPONDING MCDS

The circuit diagrams of pulse position peak responding MCDs are shown in Figure 8.9, and their associated waveforms are shown in Figure 8.10. Figure 8.9(a) shows a pulse position peak detecting MCD, and Figure 8.9(b) shows a pulse position peak sampling MCD.

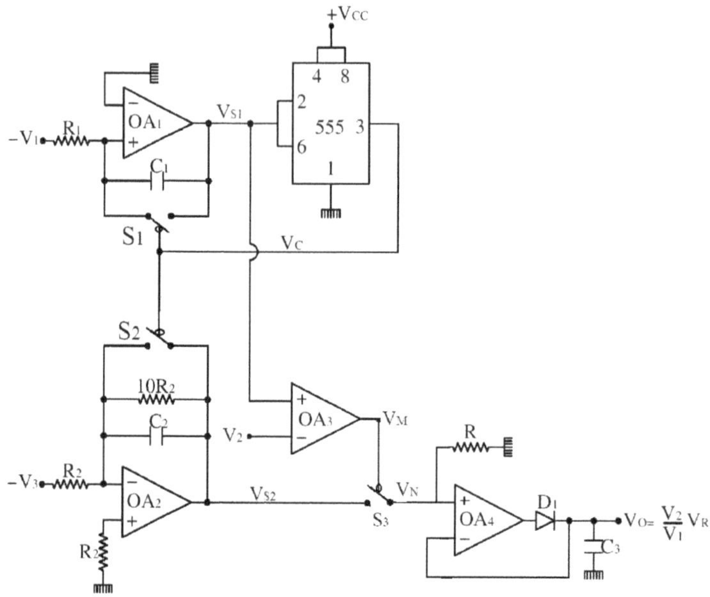

(a)

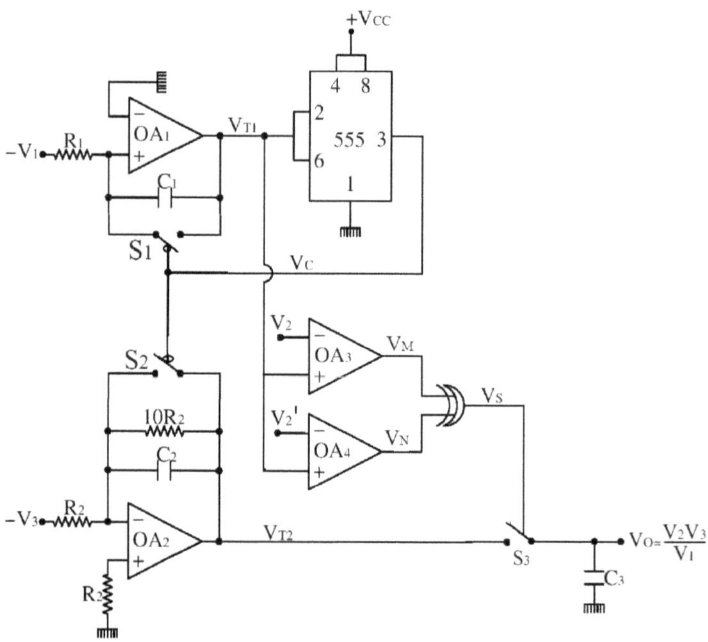

(b)

Figure 8.9 (a) Pulse position peak detecting multiplier. (b) Pulse width integrated peak sampling MCD.

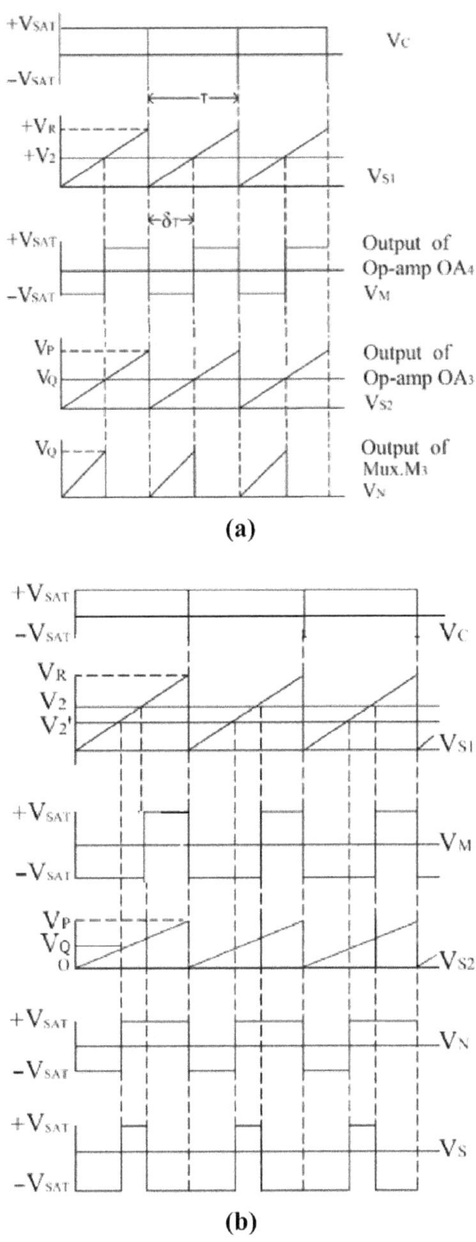

Figure 8.10 (a) Associated waveforms of Figure 8.9(a). (b) Associated waveforms of Figure 8.9(b).

In Figure 8.9(a), the integrator OA_1 and 555 timer constitute a saw tooth wave generator. The time period of the saw tooth wave V_{S1} is given as

$$T = \frac{V_R}{V_1} R_1 C_1 \tag{8.25}$$

The comparator OA_3 compares the saw tooth wave V_{S1} with an input voltage V_2 and produces a rectangular wave V_M. The OFF time δ_T of this rectangular wave V_M is given as

$$\delta_T = \frac{V_2}{V_R} T \tag{8.26}$$

The short pulse V_C in the saw tooth wave generator is also given to the switch S_2, which constitutes a controlled integrator along with the op-amp OA_2, resistor R_2, and capacitor C_2. During the HIGH value of V_C, the switch S_2 is opened, another integrator is formed by the op-amp OA_2, resistor R_2, and capacitor C_2, During the LOW value of V_C, the switch S_2 is closed, and capacitor C_2 is short circuited so that integrator OA_3 output becomes zero.
The integrator OA_3 output is given as

$$V_{S2} = -\frac{1}{R_2 C_2} \int -V_3 dt = \frac{V_3}{R_2 C_2} t \tag{8.27}$$

Another saw tooth wave V_{S2} with peak value of V_P is generated at the output of the integrator OA_3. From the waveforms shown in Figure 8.10, equation (8.27), and the fact that at t = T, $V_{S2} = V_P$,

$$V_P = \frac{V_3}{R_2 C_2} T \tag{8.28}$$

- In Figure 8.9(a), the rectangular pulse V_M controls the switch S_3. During the LOW of V_M, the switch S_3 is closed, and the saw tooth wave V_{S2} is connected to the switch S_3 output. During the HIGH of V_M, the switch S_3 is opened, and zero voltage exists on the switch S_3. A semi-saw tooth wave V_N with peak value V_Q is generated at the output of switch S_3. The peak detector realized by the op-amp OA_4, diode D_1, and capacitor C_3 gives the peak value V_Q at its output, i.e., $V_O = V_Q$.
- In Figure 8.9(b), the saw tooth wave V_{S2} is sampled by the sample and hold circuit realized by the switch S_3 and capacitor C_3 with a sampling pulse V_S. The sampling pulse V_S is generated by Ex-ORing the V_M and V_N signals. The signal V_M is generated by comparing the first saw tooth wave V_{S1} with the second input voltage V_2. The signal V_N is generated

by comparing the first saw tooth wave V_{S1} with slightly lower voltage than that of V_2, i.e., V_2'. The sample and hold output $V_O = V_Q$.

The value of V_Q is given as

$$V_Q = \frac{V_P}{T}\delta_T \tag{8.29}$$

$$V_O = \frac{V_3}{R_2C_2}\frac{V_2}{V_R}T$$

$$V_O = \frac{V_2V_3}{V_1}\frac{R_1C_1}{R_2C_2}$$

Let $R_1 = R_2$, $C_1 = C_2$.

$$V_O = \frac{V_2V_3}{V_1} \tag{8.30}$$

Chapter 9

Time Division Square Rooters (TDD)—Multiplexing

9.1 SAW TOOTH WAVE BASED TIME DIVISION SQUARE ROOTERS

The circuit diagrams of saw tooth wave based time division square rooters are shown in Figure 9.1, and their associated waveforms are shown in Figure 9.2. A saw tooth wave V_{S1} of peak value V_R and time period T is generated by the 555 timer IC.

The circuit working operation of a saw tooth wave generator is given in chapter 1.

In the circuits of Figure 9.1, the comparator OA_2 compares the saw tooth wave V_{S1} of the peak value V_R with the output voltage V_O and produces a rectangular waveform V_M at its output. The ON time δ_T of this rectangular waveform V_M is given as

$$\delta_T = \frac{V_O}{V_R} T \tag{9.1}$$

The rectangular pulse V_M controls the multiplexer M_1. When V_M is HIGH, another input voltage V_O is connected to the R_3C_2 low pass filter ('ay' is connected to 'a'). When V_M is LOW, zero voltage is connected to the R_3C_2 low pass filter ('ax' is connected to 'a'). Another rectangular pulse V_N with a maximum value of V_O is generated at the multiplexer M_1 output. The R_3C_2 low pass filter gives the average value of this pulse train V_N and is given as

$$V_X = \frac{1}{T}\int_0^{\delta_T} V_O dt \tag{9.2}$$

$$V_X = \frac{V_O}{T}\delta_T \tag{9.3}$$

DOI: 10.1201/9781003362968-9

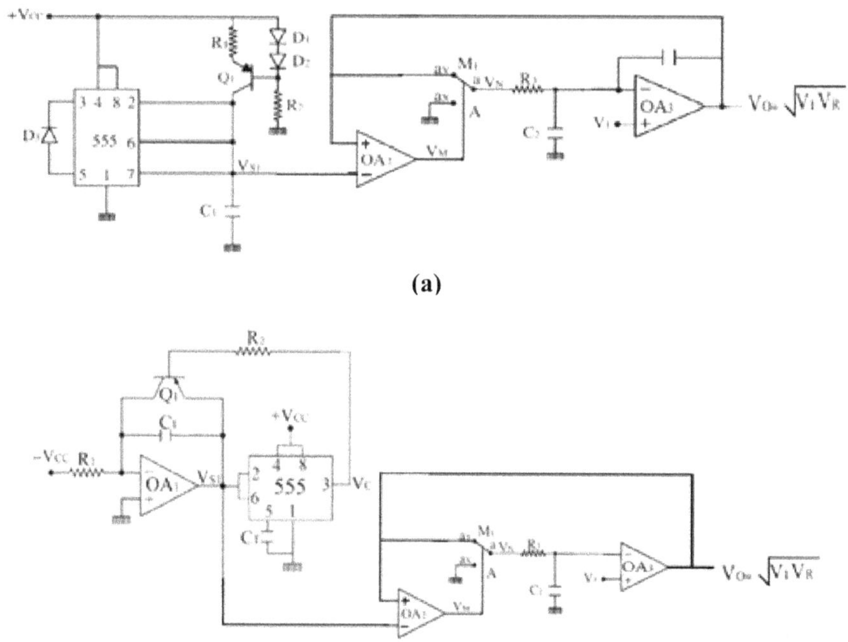

(a)

(b)

Figure 9.1 (a) Saw tooth wave based time division square rooter—type I, (b) Saw tooth wave based time division square rooter—type II.

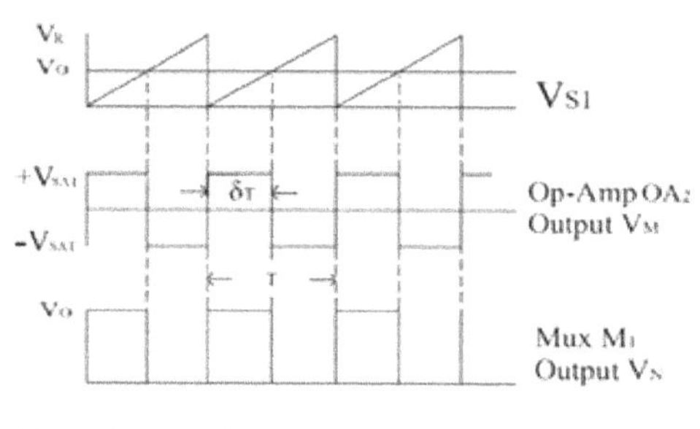

Figure 9.2 Associated waveforms of Figure 9.1.

Equation (9.1) in (9.3) gives

$$V_X = \frac{V_O^2}{V_R} \quad (9.4)$$

where $V_R = 2/3\ V_{CC}$.

The op-amp OA_3 is kept in a negative closed loop configuration, and a positive dc voltage is ensured in the feedback. Hence its inverting terminal voltage will be equal to its non-inverting terminal voltage, i.e.,

$$V_X = V_1 \tag{9.5}$$

From equations (9.4) and (9.5),

$$V_O = \sqrt{V_1 V_R} \tag{9.6}$$

9.2 TRIANGULAR WAVE REFERENCED TIME DIVISION SQUARE ROOTERS

The circuit diagrams of triangular wave based square rooters are shown in Figure 9.3, and their associated waveforms are shown in Figure 9.4. In Figure 9.3(a), a triangular wave V_{T1} with a $\pm V_T$ peak to peak value and time period T is generated by the 555 timer. The working operation of a triangular wave generator is described in chapter 1.

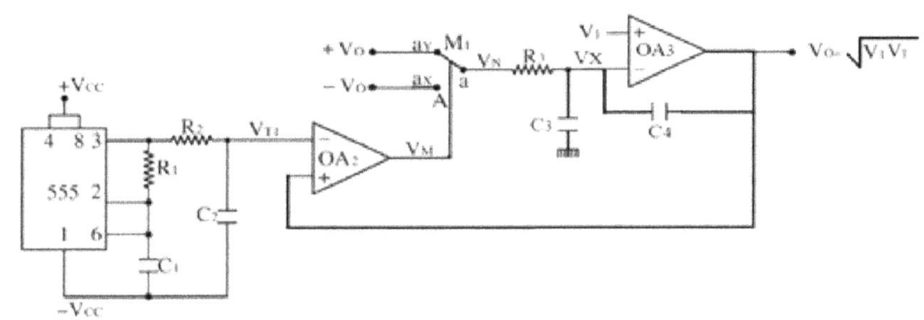

(a)

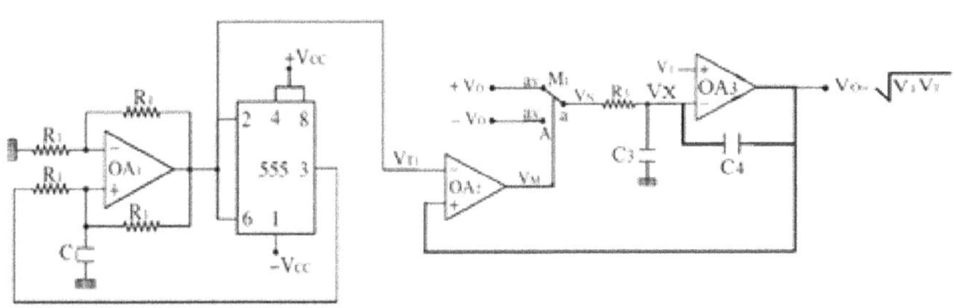

(b)

Figure 9.3 (a) Triangular wave based square rooter—type I. (b) Triangular wave based square rooter—type II.

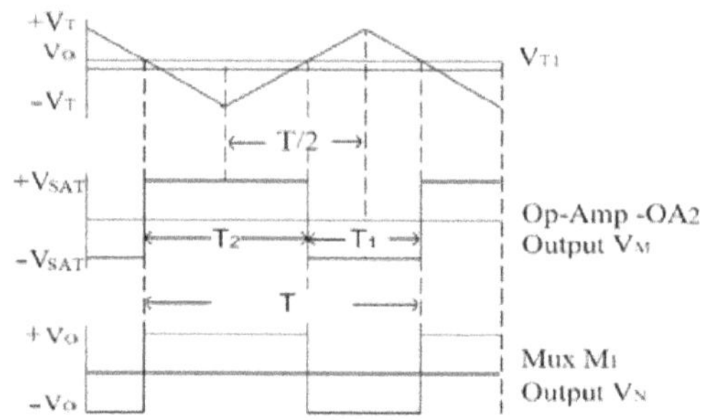

Figure 9.4 Associated waveforms of Figure 9.3(a) and (b)

In the circuits of Figure 9.3(a) and (b), the output voltage V_O is compared with the generated triangular wave V_{T1} by the comparator OA_2. An asymmetrical rectangular waveform V_M is generated at the comparator OA_2 output. From the waveforms shown in Figure 9.5, it is observed that

$$T_1 = \frac{V_T - V_O}{2V_T}T, T_2 = \frac{V_T + V_O}{2V_T}T, T = T_1 + T_2 \tag{9.7}$$

This rectangular wave V_M is given as the control input to the multiplexer M_1. The multiplexer M_1 connects the output voltage $+V_O$ during T_2 ('ay' is connected to 'a'), and $-V_O$ during T_1 ('ax' is connected to 'a'). Another rectangular asymmetrical wave V_N with a peak to peak value of $\pm V_O$ is generated at the multiplexer M_1 output. The R_3C_3 low pass filter gives the average value of the pulse train V_N, which is given as

$$V_X = \frac{1}{T}\left[\int_O^{T_2} V_O\,dt + \int_{T_2}^{T_1+T_2}(-V_O)dt\right] = \frac{V_O}{T}(T_2 - T_1) \tag{9.8}$$

Equation (9.7) in (9.8) gives

$$V_X = \frac{V_O^{\ 2}}{V_T} \tag{9.9}$$

where $V_T = V_{CC}/3$. (9.10)

The op-amp OA_3 is kept in a negative closed loop configuration, and a positive dc voltage is ensured in the feedback. Hence its inverting terminal voltage will be equal to its non-inverting terminal voltage, i.e.,

$$V_X = V_1 \tag{9.11}$$

From equations (9.9) and (9.11),

$$V_O = \sqrt{V_1 V_T} \tag{9.12}$$

9.3 TIME DIVISION SQUARE ROOTER WITH NO REFERENCE—TYPE I

The square rooter using the time division principle without using any reference clock is shown in Figure 9.5, and its associated waveforms are shown in Figure 9.6.

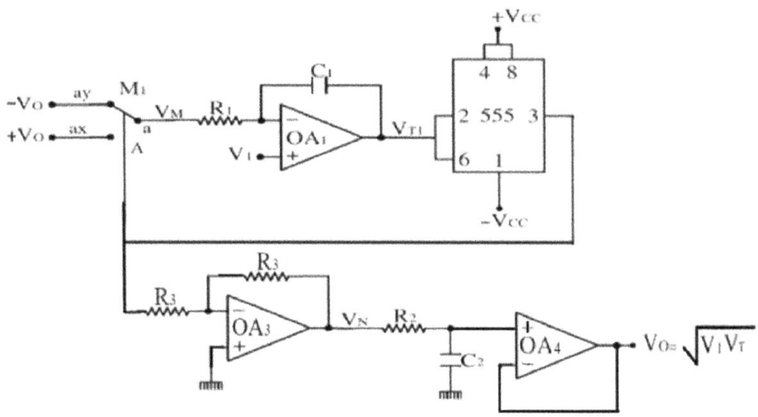

Figure 9.5 Time division square rooter without reference clock.

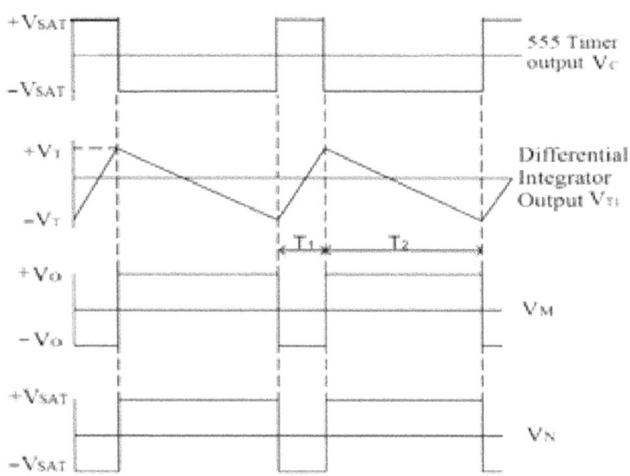

Figure 9.6 Associated waveforms of Figure 9.5.

Initially the 555 timer output is HIGH. The multiplexer M_1 connects $-V_O$ to the differential integrator composed by resistor R_1, capacitor C_1, and op-amp OA_1 ('ay' is connected to 'a'). The output of the differential integrator will be

$$V_{T1} = \frac{1}{R_1 C_1}\int (V_1 + V_O)dt$$

$$V_{T1} = \frac{(V_O + V_1)}{R_1 C_1}t \tag{9.13}$$

The output of the differential integrator rises toward positive saturation, and when it reaches the voltage level of $+V_T$, the 555 timer output becomes LOW. The multiplexer M_1 connects $+V_O$ to the differential integrator composed of the resistor R_1, capacitor C_1, and op-amp OA_1 ('ax' is connected to 'a'). Now the output of differential integrator will be

$$V_{T1} = \frac{1}{R_1 C_1}\int (V_1 - V_O)dt$$

$$V_{T1} = -\frac{(V_O - V_1)}{R_1 C_1}t \tag{9.14}$$

The output of the differential integrator reverses toward negative saturation, and when it reaches the voltage level $-V_T$, the 555 timer output becomes HIGH and the cycle therefore repeats, to give an asymmetrical rectangular wave V_C at the output of the 555 timer.

$$V_T = \frac{V_{CC}}{3} \tag{9.15}$$

From the waveforms shown in Figure 9.6, it is observed that

$$T_1 = \frac{V_O - V_1}{2V_O}T, T_2 = \frac{V_O + V_1}{2V_O}T, T = T_1 + T_2 \tag{9.16}$$

Another rectangular wave V_N is generated at the output of the inverting amplifier OA_3. The R_2C_2 low pass filter gives the average value of this pulse train V_N and is given as

$$V_O = \frac{1}{T}\left[\int_O^{T_2} V_{SAT}\,dt + \int_{T_2}^{T_1+T_2} (-V_{SAT})dt\right]$$

$$V_O = \frac{V_{SAT}(T_2 - T_1)}{T} \tag{9.17}$$

Equation (9.16) in (9.17) gives

$$V_O = \sqrt{V_1 V_{SAT}} \tag{9.18}$$

9.4 TIME DIVISION SQUARE ROOTER WITH NO REFERENCE—TYPE II

The square rooter using the time division principle without using any reference clock is shown in Figure 9.7, and its associated waveforms are shown in Figure 9.8.

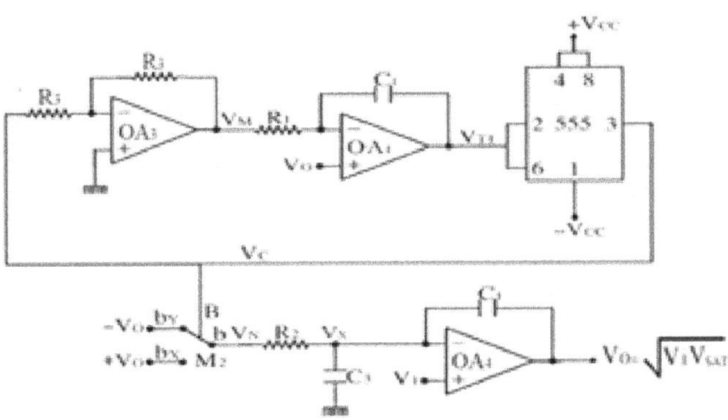

Figure 9.7 Square rooter without reference clock—type II.

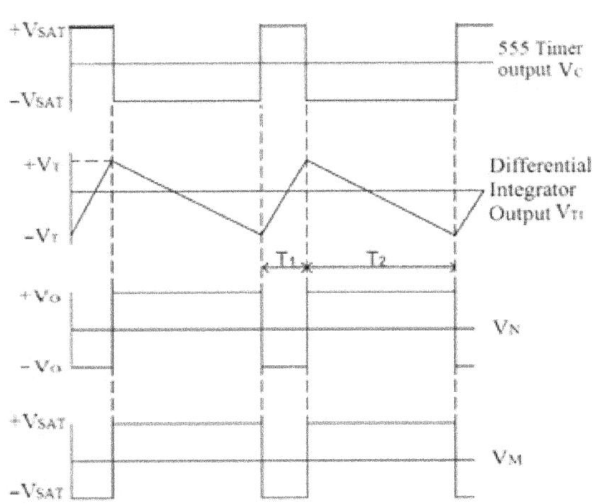

Figure 9.8 Associated waveforms of Figure 9.7.

Initially the 555 timer output is HIGH. The inverting amplifier OA_3 output will be $-V_{SAT}$. The output of differential integrator will be

$$V_{T1} = \frac{1}{R_1C_1}\int (V_O + V_{SAT})dt$$

$$V_{T1} = \frac{(V_O + V_{SAT})}{R_1C_1}t \tag{9.19}$$

The output of the differential integrator is rising toward positive saturation, and when it reaches the voltage level of $+V_T$, the 555 timer output becomes LOW. The inverting amplifier OA_3 output will be $+V_{SAT}$. Now the output of the differential integrator will be

$$V_{T1} = \frac{1}{R_1C_1}\int (V_O - V_{SAT})dt$$

$$V_{T1} = -\frac{(V_{SAT} - V_O)}{R_1C_1}t \tag{9.20}$$

The output of the differential integrator reverses toward negative saturation, and when it reaches the voltage level $-V_T$, the 555 timer output becomes HIGH, and the cycle therefore repeats, to give an asymmetrical rectangular wave V_C at the output of the 555 timer.

$$V_T = \frac{V_{CC}}{3} \tag{9.21}$$

From the waveforms shown in Figure 9.8, it is observed that

$$T_1 = \frac{V_{SAT} - V_O}{2V_{SAT}}T, T_2 = \frac{V_{SAT} + V_O}{2V_{SAT}}T, T = T_1 + T_2 \tag{9.22}$$

The asymmetrical rectangular wave V_C controls multiplexer M_2. The multiplexer M_2 connects $-V_O$ during the ON time T_2 ('by' is connected to 'b') and $+V_O$ during the OFF time T_1 of the rectangular wave V_C ('bx' is connected to 'b'). Another rectangular wave V_N with $\pm V_O$ as the peak to peak value is generated at the multiplexer M_2 output. The R_2C_2 low pass filter gives the average value of this pulse train V_N and is given as

$$V_X = \frac{1}{T}\left[\int_O^{T_2} V_O\, dt + \int_{T_2}^{T_1+T_2} (-V_O)dt\right]$$

$$V_X = \frac{V_O(T_2 - T_1)}{T} \tag{9.23}$$

Equation (9.22) in (9.23) gives

$$V_X = \frac{V_O^{\ 2}}{V_{SAT}} \tag{9.24}$$

The op-amp OA_4 is at a negative closed loop configuration, and a positive dc voltage is ensured in the feedback loop. Hence its non-inverting terminal voltage is equal to its inverting terminal voltage, i.e.,

$$V_1 = V_X \tag{9.25}$$

From equations (9.24) and (9.25),

$$V_O = \sqrt{V_1 V_{SAT}} \tag{9.26}$$

9.5 SQUARE ROOTER FROM 555 ASTABLE MULTIVIBRATOR

9.5.1 Type I

The circuit diagram of a square rooter using the 555 astable multivibrator is shown in Figure 9.9, and its associated waveforms are shown in Figure 9.10. Refer to the internal diagram of the 555 timer IC shown in Figure 0.1. Initially when we switch on the power supply, the output of the upper comparator CMP_1 will be LOW, i.e., R = 0, and the output of the lower comparator CMP_2 will be HIGH, i.e., S = 1. The flip flop outputs are Q = 1 and Q′ = 0. The timer output at pin 3 will be HIGH, transistor Q_1 is OFF, and hence the discharge pin 7 is at the open position.

The capacitor C_1 is charging toward V_O through the resistors R_1 and R_2 with a time constant of $(R_1+R_2)C_1$, and its voltage is rising exponentially. When the capacitor voltage is rising above the voltage 2/3 V_{CC}, the output of the upper comparator CMP_1 becomes HIGH, i.e., R = 1, and the output of

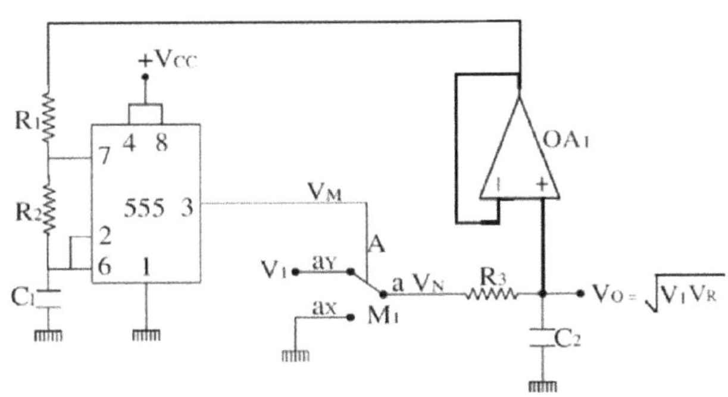

Figure 9.9 Square rooter from 555 astable.

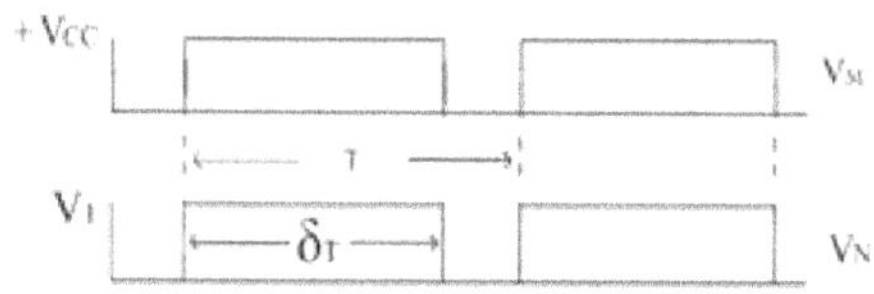

Figure 9.10 Associated waveforms of Figure 9.9.

the lower comparator CMP_2 becomes LOW, i.e., S = 0. The flip flop outputs are Q = 0 and Q′ = 1. The timer output at pin 3 will be LOW, transistor Q_1 is ON, and hence the discharge pin 7 is at GND potential. Now the capacitor C_1 is discharging to GND potential through the resistor R_2 with a time constant of R_2C. When the capacitor voltage falls below 1/3 V_{CC}, the output of the upper comparator CMP_1 becomes LOW, i.e., R = 0, and the output of the lower comparator CMP_2 becomes HIGH, i.e., S = 1. The flip flop outputs are Q = 1 and Q′ = 0. The timer output at pin 3 will be HIGH, transistor Q_1 is OFF, and hence the discharge pin 7 is at the open position.

Now the capacitor starts charging toward V_O, and the cycle therefore repeats to produce periodic pulses at the output pin 3 of the 555 timer.

The ON time of the 555 timer output V_M is inversely proportional to V_O. During the ON time δ_T, the input voltage V_1 is connected to the R_3C_2 low pass filter ('ay' is connected to 'a'). During the OFF time of V_M, zero voltage is connected to the R_3C_2 low pass filter ('ax' is connected to 'a'). Another rectangular waveform V_N with V_1 as the peak value is generated at the output of the multiplexer M_1. The ON time δ_T of this rectangular pulse V_N is given as

$$\delta_T = \frac{V_R}{V_O} T \tag{9.27}$$

where V_R is a constant value.

The R_3C_2 low pass filter gives the average value of this pulse train V_N and is given as

$$V_O = \frac{1}{T}\int_0^{\delta_T} V_1 dt = \frac{V_1}{T}\delta_T$$

$$V_O = \sqrt{V_1 V_R} \tag{9.28}$$

9.5.2 Square Rooter from 555 Astable Multivibrator—Type II

The circuit diagram of the square rooter using the 555 astable multivibrator is shown in Figure 9.11, and its associated waveforms are shown in Figure 9.12. Refer to the internal diagram of the 555 timer IC shown in

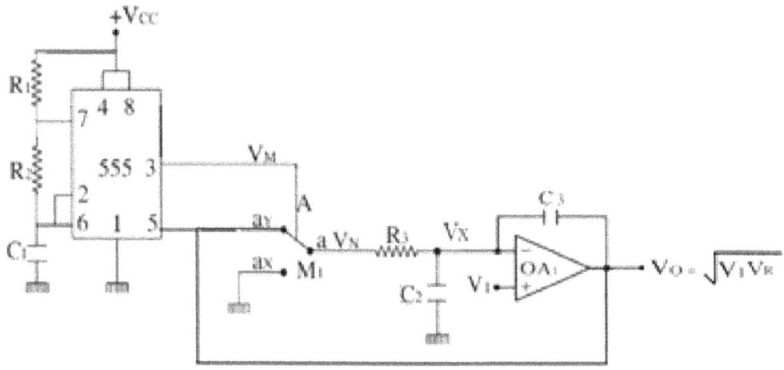

Figure 9.11 Square rooter with the 555 timer astable multivibrator.

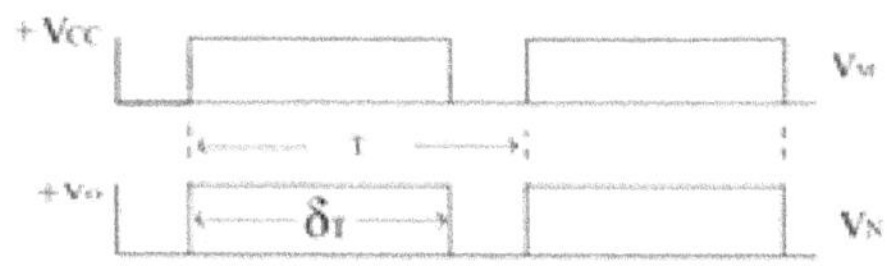

Figure 9.12 Associated waveforms of Figure 9.11.

Figure 0.1. Initially when we switch on the power supply, the output of the upper comparator CMP_1 will be LOW, i.e., R = 0, and the output of the lower comparator CMP_2 will be HIGH, i.e., S = 1. The flip flop outputs are Q = 1 and Q′ = 0. The timer output at pin 3 will be HIGH, transistor Q_1 is OFF, and hence the discharge pin 7 is at the open position.

The capacitor C_1 is charging toward $+V_{CC}$ through the resistors R_1 and R_2 with a time constant of $(R_1+R_2)C_1$, and its voltage is rising exponentially. When the capacitor voltage is rising above the voltage V_O, the output of the upper comparator CMP_1 becomes HIGH, i.e., R = 1, and the output of the lower comparator CMP_2 becomes LOW, i.e., S = 0. The flip flop outputs are Q = 0 and Q′ = 1. The timer output at pin 3 will be LOW, transistor Q_1 is ON, and hence the discharge pin 7 is at GND potential. Now the capacitor C_1 is discharging to GND potential through the resistor R_2 with a time constant of R_2C. When the capacitor voltage falls below 1/3 V_{CC}, the output of the upper comparator CMP_1 becomes LOW, i.e., R = 0, and the output of the lower comparator CMP_2 becomes HIGH, i.e., S = 1. The flip flop outputs are Q = 1 and Q′ = 0. The timer output at pin 3 will be HIGH, transistor Q_1 is OFF, and hence the discharge pin 7 is at the open position.

Now the capacitor starts charging toward $+V_{CC}$, and the cycle therefore repeats to produce periodic pulses at the output pin 3 of the 555 timer.

The ON time of the 555 timer output V_M is proportional to V_O, which is applied at its pin 5. During the ON time δ_T, the output voltage V_O is connected to the R_3C_2 low pass filter ('ay' is connected to 'a'). During the OFF time of V_M, zero voltage is connected to the R_3C_2 low pass filter ('ax' is connected to 'a'). Another rectangular waveform V_N with V_O as a peak value is generated at the output of multiplexer M_1.

$$\delta_T = \frac{V_O}{V_R} T \tag{9.29}$$

The R_3C_2 low pass filter gives the average value of this pulse train V_N and is given as

$$V_X = \frac{1}{T}\int_0^{\delta_T} V_O dt = \frac{V_O}{T}\delta_T$$

$$V_X = \frac{V_O^2}{V_R} \tag{9.30}$$

where V_R is a constant value.

The op-amp OA_1 is kept in a negative closed loop configuration, and a positive dc voltage is ensured in the feedback. Hence its inverting terminal voltage will be equal to its non-inverting terminal voltage, i.e.,

$$V_X = V_1 \tag{9.31}$$

From equations (9.30) and (9.31),

$$V_O = \sqrt{V_1 V_R} \tag{9.32}$$

9.6 SQUARE ROOTER FROM 555 MONOSTABLE MULTIVIBRATOR

9.6.1 Type I

The circuit diagrams of a square rooter using the 555 monostable multivibrator are shown in Figure 9.13, and its associated waveforms are shown in Figure 9.14. Refer to the internal diagram of the 555 timer IC shown in Figure 0.1. Initially when the power supply is switch on, the output of the upper comparator CMP_1 will be LOW, i.e., R = 0, and the output of the lower comparator CMP_2 will be HIGH, i.e., S = 1. The flip flop outputs are Q = 1 and Q′ = 0. The timer output at pin 3 will be HIGH, transistor Q_1 is OFF, and hence the discharge pin 7 is at the open position. The capacitor C_1 is charging toward V_O through the resistor R_1. The capacitor voltage is rising exponentially, and when it reaches the value of 2/3 V_{CC}, the output of the

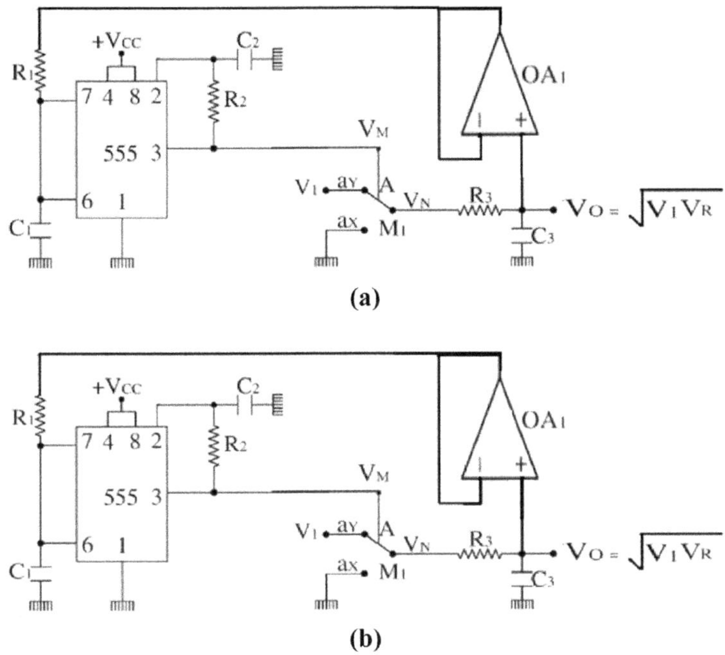

Figure 9.13 (a) Square rooter using 555 timer monostable multivibrator. (b) Square rooter using re-trigger monostable multivibrator.

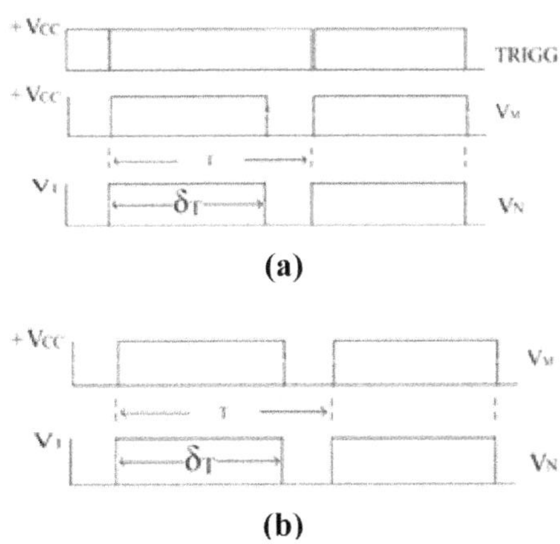

Figure 9.14 (a) associated waveforms of Figure 9.13(a). (b) Associated waveforms of Figure 9.13(b).

upper comparator CMP_1 becomes HIGH, i.e., R = 1, and the output of the lower comparator CMP_2 becomes LOW, i.e., S = 0. The flip flop outputs are Q = 0 and Q′ = 1. The timer output at pin 3 will be LOW, transistor Q_1 is ON, and hence the discharge pin 7 is at GND potential. Now the capacitor C_1 is short circuited, zero voltage exists at pin 6, and the output of the upper comparator CMP_1 becomes LOW, i.e., R = 0. A trigger pulse is applied at pin 2 and when the trigger voltage comes down to 1/3 V_{CC}, the output of the lower comparator CMP_2 becomes HIGH, i.e., S = 1. The flip flop outputs are Q = 1 and Q′ = 0. The timer output at pin 3 will be HIGH, transistor Q_1 is OFF, and hence the discharge pin 7 is at the open position.

Now the capacitor C_1 is charging toward V_O, and the sequence therefore repeats for every trigger input pulse.

The ON time of the 555 timer output V_M is inversely proportional to V_O. The output of the 555 timer controls the multiplexer M_1. During the ON time δ_T, the second input voltage V_1 is connected to the R_3C_3 low pass filter ('ay' is connected to 'a'). During the OFF time of V_M, zero voltage is connected to the R_3C_3 low pass filter ('ax' is connected to 'a'). Another rectangular waveform V_N with V_1 as the peak value is generated at the output of the multiplexer M_1.

$$\delta_T = \frac{V_R}{V_O} T \tag{9.33}$$

The R_3C_3 low pass filter gives the average value of this pulse train V_N and is given as

$$V_O = \frac{1}{T}\int_0^{\delta_T} V_1 dt = \frac{V_1}{T}\delta_T$$

$$V_O = \sqrt{V_1 V_R} \tag{9.34}$$

where V_R is a constant value.

Figure 9.13(b) shows re-trigger monostable multivibrator used as analog square rooter.

9.6.2 Square Rooter from 555 Monostable Multivibrator—Type II

The circuit diagrams of a square rooter using the 555 timer monostable multivibrator are shown in Figure 9.15, and their associated waveforms are shown in Figure 9.16. Refer to the internal diagram of the 555 timer IC shown in Figure 0.1. Initially when the power supply is switched on, the output of the upper comparator CMP_1 will be LOW, i.e., R = 0, and the

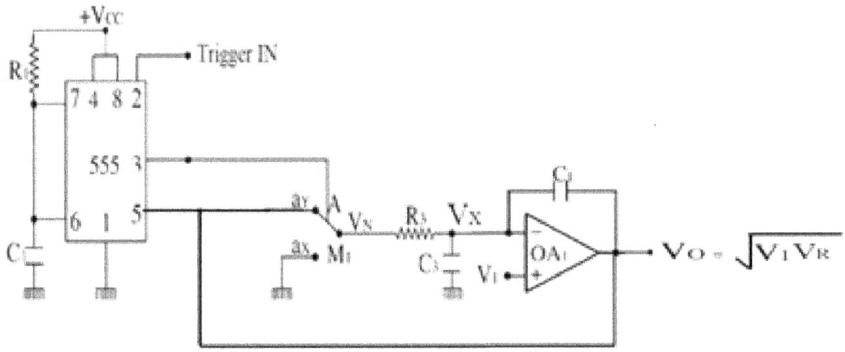

(a)

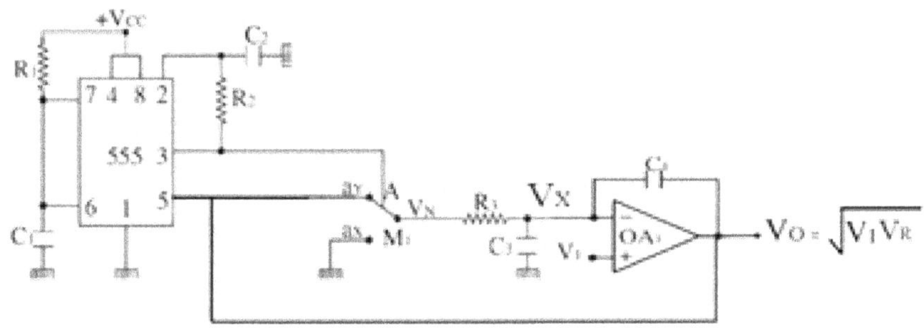

(b)

Figure 9.15 (a) Square rooter from 555 monostable. (b) Square rooter with 555 re-trigger monostable multivibrator.

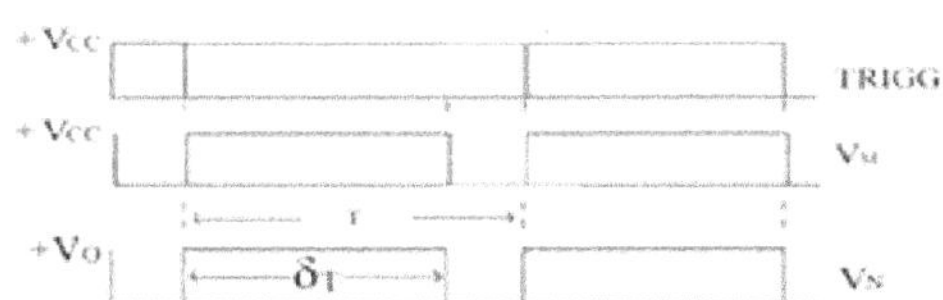

(a)

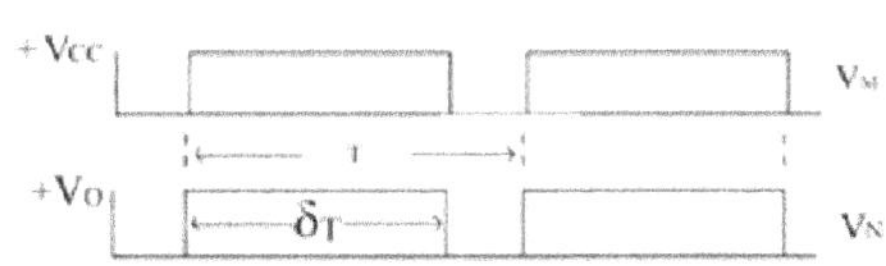

(b)

Figure 9.16 (a) Associated waveforms of Figure 9.15(a). (b) Associated waveforms of Figure 9.15(b).

output of the lower comparator CMP_2 will be HIGH, i.e., S = 1. The flip flop outputs are Q = 1 and Q′ = 0. The timer output at pin 3 will be HIGH, transistor Q_1 is OFF, and hence the discharge pin 7 is at the open position. The capacitor C_1 is charging toward $+V_{CC}$ through the resistor R_1. The capacitor voltage is rising exponentially, and when it reaches the value of V_O, the output of the upper comparator CMP_1 becomes HIGH, i.e., R = 1, and the output of the lower comparator CMP_2 becomes LOW, i.e., S = 0. The flip flop outputs are Q = 0 and Q′ = 1. The timer output at pin 3 will be LOW, transistor Q_1 is ON, and hence the discharge pin 7 is at GND potential. Now the capacitor C_1 is short circuited, zero voltage exists at pin 6, the output of the upper comparator CMP_1 becomes LOW, i.e., R = 0. A trigger pulse is applied at pin 2, and when the trigger voltage comes down to 1/3 V_{CC}, the output of the lower comparator CMP_2 becomes HIGH, i.e., S = 1. The flip flop outputs are Q = 1 and Q′ = 0. The timer output at pin 3 will be HIGH, transistor Q_1 is OFF, and hence the discharge pin 7 is at the open position.

Now the capacitor C_1 is charging toward $+V_{CC}$, and the sequence therefore repeats for every trigger input pulse.

The ON time of the 555 timer output V_M is proportional to V_O, which is applied at its pin 5. The 555 timer output controls the multiplexer M_1. During the ON time δ_T, the output voltage V_O is connected to the R_3C_2 low pass filter ('ay' is connected to 'a'). During the OFF time of V_M, zero voltage is connected to the R_3C_2 low pass filter ('ax' is connected to 'a'). Another rectangular waveform V_N with V_O as a peak value is generated at the output of multiplexer M_1. The ON time δ_T of this rectangular waveform V_N is given as

$$\delta_T = \frac{V_O}{V_R} T \tag{9.35}$$

The R_3C_3 low pass filter gives the average value of this pulse train V_N and is given as

$$V_X = \frac{1}{T}\int_0^{\delta_T} V_O dt = \frac{V_O}{T}\delta_T$$

$$V_X = \frac{V_O^{\,2}}{V_R} \tag{9.36}$$

where V_R is a constant value.

The op-amp OA_1 is kept in a negative closed loop configuration, and a positive dc voltage is ensured in the feedback. Hence its inverting terminal voltage will be equal to its non-inverting terminal voltage, i.e.,

$$V_X = V_1 \tag{9.37}$$

From equations (9.36) and (9.37),

$$V_O = \sqrt{V_1 V_R} \tag{9.38}$$

Figure 9.15(b) shows a re-trigger monostable as a square rooter.

Chapter 10

Time Division Square Rooters (TDSR)—Switching

10.1 SAW TOOTH WAVE BASED TIME DIVISION SQUARE ROOTERS

The circuit diagrams of saw tooth wave based time division square rooters are shown in Figure 10.1, and their associated waveforms are shown in Figure 10.2. A saw tooth wave V_{S1} of peak value V_R and time period T is generated by the 555 timer.

In the circuit of Figure 10.1, the comparator OA_2 compares the saw tooth wave V_{S1} of the peak value V_R with the output voltage V_O and produces a rectangular waveform V_M at its output. The ON time δ_T of this rectangular waveform V_M is given as

$$\delta_T = \frac{V_O}{V_R} T \tag{10.1}$$

The rectangular pulse V_M controls the switch S_1. When V_M is HIGH, another input voltage V_O is connected to the R_3C_2 low pass filter (switch S_1 is closed). When V_M is LOW, zero voltage is connected to the R_3C_2 low pass filter (switch S_1 is opened). Another rectangular pulse V_N with a maximum value of V_2 is generated at the switch S_1 output. The R_3C_2 low pass filter gives the average value of this pulse train V_N and is given as

$$V_X = \frac{1}{T}\int_0^{\delta_T} V_O dt \tag{10.2}$$

$$V_X = \frac{V_O}{T}\delta_T \tag{10.3}$$

Equation (10.1) in (10.3) gives

$$V_X = \frac{V_O^{\ 2}}{V_R} \tag{10.4}$$

DOI: 10.1201/9781003362968-10

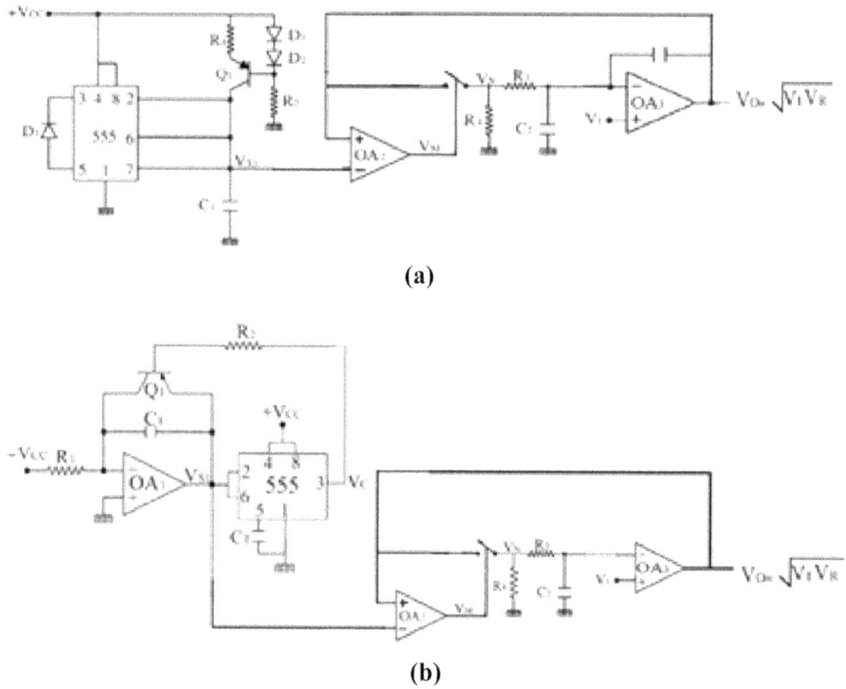

Figure 10.1 (a) Saw tooth wave based time division square rooter—type I. (b) Saw tooth wave based time division square rooter—type II.

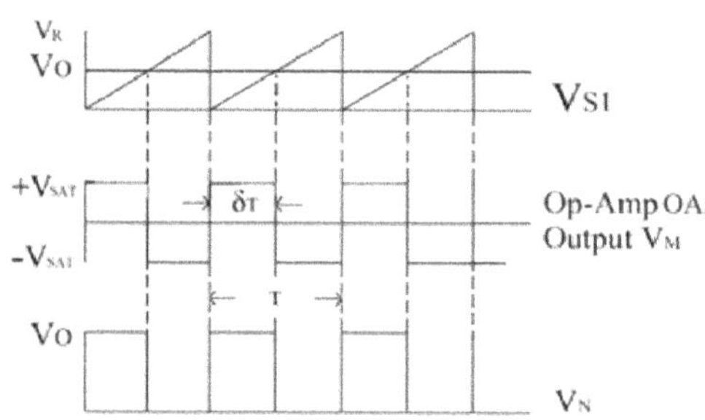

Figure 10.2 Associated waveforms of Figure 10.1.

where $V_R = 2/3\ V_{CC}$.

The op-amp OA_3 is kept in a negative closed loop configuration, and a positive dc voltage is ensured in the feedback. Hence its inverting terminal voltage will be equal to its non-inverting terminal voltage, i.e.,

$$V_X = V_1 \tag{10.5}$$

From equations (10.4) and (10.5),

$$V_O = \sqrt{V_1 V_R} \tag{10.6}$$

10.2 TRIANGULAR WAVE REFERENCED TIME DIVISION SQUARE ROOTERS

The circuit diagrams of triangular wave based square rooters are shown in Figure 10.3, and their associated waveforms are shown in Figure 10.4. A triangular wave V_{T1} with a $\pm V_T$ peak to peak value and time period T is generated by the 555 timer. The output voltage V_O is compared with the generated triangular wave V_{T1} by the comparator on OA_2. An asymmetrical rectangular waveform V_M is generated at the comparator OA_2 output. From the waveforms shown in Figure 10.4, it is observed that

$$T_1 = \frac{V_T - V_O}{2V_T} T \quad T_2 = \frac{V_T + V_O}{2V_T} T \quad T = T_1 + T_2 \tag{10.7}$$

This rectangular wave V_M is given as the control input to the switch S_1. During T_2 of V_M, the switch S_1 is closed, and the op-amp OA_3 will work as a non-inverting amplifier. $+V_O$ will be its output, i.e., $V_N = +V_O$. During

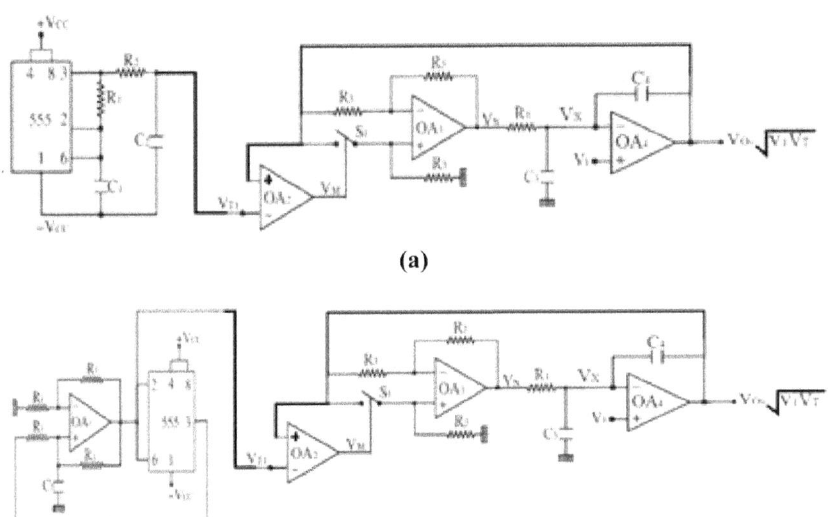

Figure 10.3 (a) Triangular wave based square rooter—type I. (b) Triangular wave based square rooter—type II.

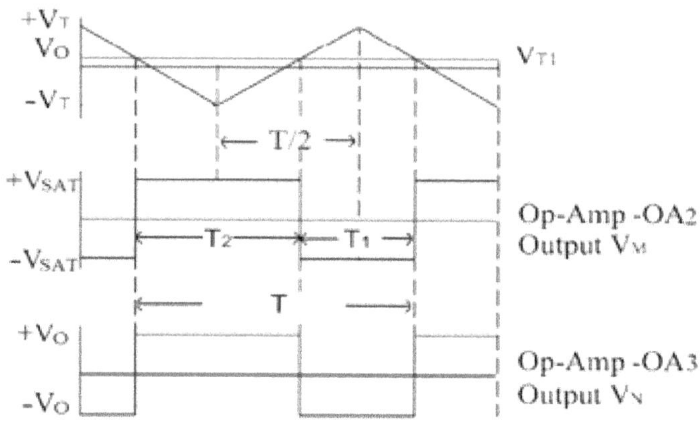

Figure 10.4 Associated waveforms of Figure 10.3(a) and (b).

T_1 of V_M, the switch S_1 is opened, and the op-amp will work as an inverting amplifier. $-V_O$ will be at its output, i.e., $V_N = -V_O$. Another rectangular asymmetrical wave V_N with a peak to peak value of $\pm V_O$ is generated at the op-amp OA_3 output. The R_4C_3 low pass filter gives the average value of the pulse train V_N and is given as

$$V_X = \frac{1}{T}\left[\int_O^{T_2} V_O\,dt + \int_{T_2}^{T_1+T_2} (-V_O)\,dt\right] = \frac{V_O}{T}(T_2 - T_1) \tag{10.8}$$

Equation (10.7) in (10.8) gives

$$V_X = \frac{V_O^{\ 2}}{V_T} \tag{10.9}$$

where $V_T = V_{CC}/3$. (10.10)

The op-amp OA_4 is kept in a negative closed loop configuration, and a positive dc voltage is ensured in the feedback. Hence its inverting terminal voltage will be equal to its non-inverting terminal voltage, i.e.,

$$V_X = V_1 \tag{10.11}$$

From equations (10.9) and (10.11),

$$V_O = \sqrt{V_1 V_T} \tag{10.12}$$

10.3 TIME DIVISION SQUARE ROOTER WITH NO REFERENCE—TYPE I

The square rooter using the time division principle without using any reference clock is shown in Figure 10.5, and its associated waveforms are shown in Figure 10.6. Initially the 555 timer output is HIGH. The switch S_1 is closed, and the op-amp OA_3 will work as a non-inverting amplifier. $-V_O$ is given to the differential integrator composed of the resistor R_1, capacitor C_1, and op-amp OA_1. The output of differential integrator will be

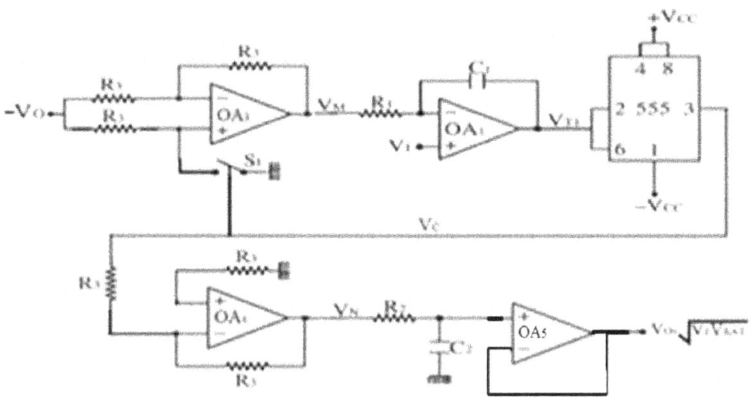

Figure 10.5 Time division square rooter without reference clock.

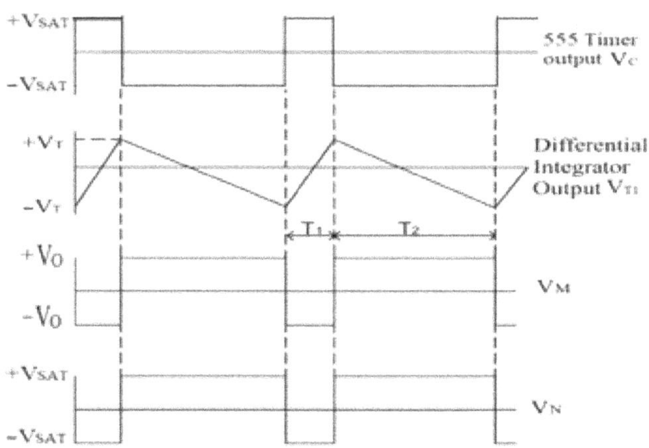

Figure 10.6 Associated waveforms of Figure 10.5.

$$V_{T1} = \frac{1}{R_1C_1}\int (V_1 + V_O)dt$$

$$V_{T1} = \frac{(V_O + V_1)}{R_1C_1}t \tag{10.13}$$

The output of the differential integrator is rising toward positive saturation, and when it reaches the voltage level of $+V_T$, the 555 timer output becomes LOW. The switch S_1 is opened, and the op-amp OA_3 will work as an inverting amplifier. $+V_O$ is given to the differential integrator composed of the resistor R_1, capacitor C_1, and op-amp OA_1. Now the output of the differential integrator will be

$$V_{T1} = \frac{1}{R_1C_1}\int (V_1 - V_O)dt$$

$$V_{T1} = -\frac{(V_O - V_1)}{R_1C_1}t \tag{10.14}$$

The output of the differential integrator reverses toward negative saturation, and when it reaches the voltage level $-V_T$, the 555 timer output becomes HIGH, and the cycle therefore repeats, to give an asymmetrical rectangular wave V_C at the output of the 555 timer.

$$V_T = \frac{V_{CC}}{3} \tag{10.15}$$

From the waveforms shown in Figure 10.6, it is observed that

$$T_1 = \frac{V_O - V_1}{2V_O}T, T_2 = \frac{V_O + V_1}{2V_O}T, T = T_1 + T_2 \tag{10.16}$$

The 555 timer output is given to the inverting amplifier OA_4. Another rectangular wave V_N is generated at the op-amp OA_4 output with $\pm V_{SAT}$ peak to peak values. The R_2C_2 low pass filter gives the average value of this pulse train V_N and is given as

$$V_O = \frac{1}{T}\left[\int_O^{T_2} V_{SAT}\,dt + \int_{T_2}^{T_1+T_2} (-V_{SAT})dt\right]$$

$$V_O = \frac{V_{SAT}(T_2 - T_1)}{T} \tag{10.17}$$

Equation (10.16) in (10.17) gives

$$V_O = \sqrt{V_1 V_{SAT}} \tag{10.18}$$

10.4 TIME DIVISION SQUARE ROOTER WITH NO REFERENCE—TYPE II

The square rooter using the time division principle without using any reference clock is shown in Figure 10.7, and its associated waveforms are shown in Figure 10.8.

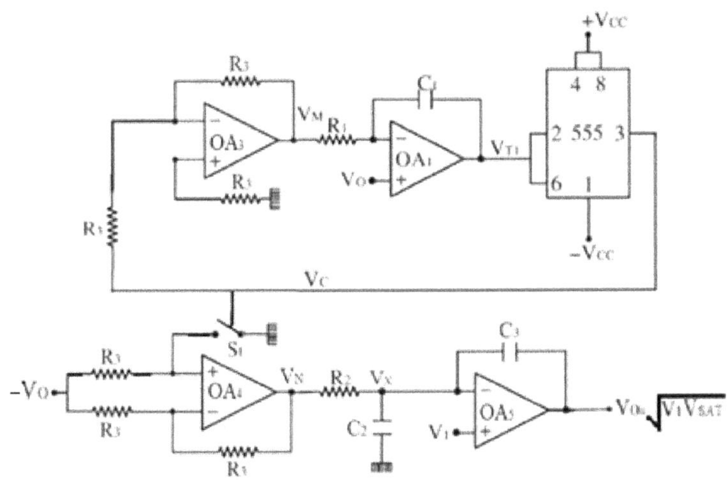

Figure 10.7 Square rooter without reference clock—type II.

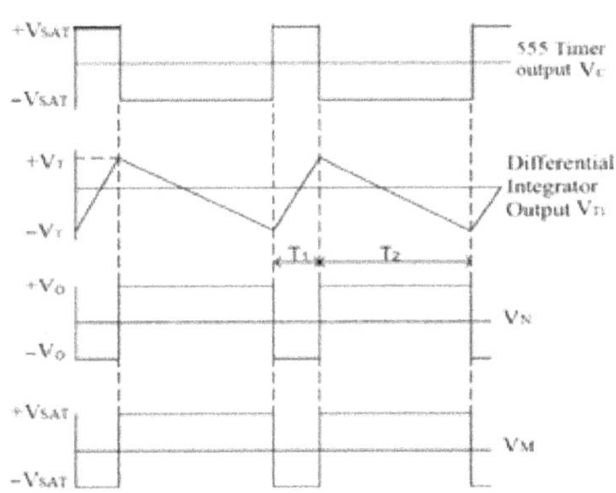

Figure 10.8 Associated waveforms of Figure 10.7.

Initially the 555 timer output is HIGH. The inverting amplifier OA_3 output will be LOW. The output of differential integrator will be

$$V_{T1} = \frac{1}{R_1C_1}\int (V_O + V_{SAT})dt$$

$$V_{T1} = \frac{(V_O + V_{SAT})}{R_1C_1}t \tag{10.19}$$

The output of the differential integrator is rising toward positive saturation, and when it reaches the voltage level of $+V_T$, the 555 timer output becomes LOW. The inverting amplifier OA_3 output will be HIGH. The output of differential integrator will now be

$$V_{T1} = \frac{1}{R_1C_1}\int (V_O - V_{SAT})dt$$

$$V_{T1} = -\frac{(V_{SAT} - V_O)}{R_1C_1}t \tag{10.20}$$

The output of the differential integrator reverses toward negative saturation, and when it reaches the voltage level $-V_T$, the 555 timer output becomes HIGH, and the cycle therefore repeats, to give an asymmetrical rectangular wave V_C at the output of the 555 timer.

$$V_T = \frac{V_{CC}}{3} \tag{10.21}$$

From the waveforms shown in Figure 10.8, it is observed that

$$T_1 = \frac{V_{SAT} - V_O}{2V_{SAT}}T, T_2 = \frac{V_{SAT} + V_O}{2V_{SAT}}T, T = T_1 + T_2 \tag{10.22}$$

During the HIGH of V_C, the S_1 is closed, the op-amp OA_4 will work as a non-inverting amplifier, and $-V_O$ is given to the low pass filter. During the LOW of V_C, the S_1 is opened, the op-amp OA_2 will work as an inverting amplifier, and $+V_O$ is given to a low pass filter. Another rectangular wave V_N with $\pm V_O$ as the peak to peak value is generated at the output of the op-amp OA_4. The R_2C_2 low pass filter gives the average value of this pulse train V_N and is given as

$$V_X = \frac{1}{T}\left[\int_{O}^{T_2} V_O\,dt + \int_{T_2}^{T_1+T_2} (-V_O)\,dt\right]$$

$$V_X = \frac{V_O(T_2 - T_1)}{T} \tag{10.23}$$

Equation (10.22) in (10.23) gives

$$V_X = \frac{V_O^{\ 2}}{V_{SAT}} \tag{10.24}$$

The op-amp OA_5 is in a negative closed loop configuration, and a positive dc voltage is ensured in the feedback loop. Hence its non-inverting terminal voltage is equal to its inverting terminal voltage, i.e.,

$$V_1 = V_X \tag{10.25}$$

From equations (10.24) and (10.25),

$$V_O = \sqrt{V_1 V_{SAT}} \tag{10.26}$$

10.5 SQUARE ROOTER FROM 555 ASTABLE MULTIVIBRATOR

10.5.1 Type I

The circuit diagram of square rooter using the 555 astable multivibrator is shown in Figure 10.9, and its associated waveforms are shown in Figure 10.10. Refer to the internal diagram of the 555 timer IC shown in Figure 0.1. Initially when we switch on the power supply, the output of the upper comparator CMP_1 will be LOW, i.e., R = 0, and the output of the lower comparator CMP_2 will be HIGH, i.e., S = 1. The flip flop outputs are Q = 1 and Q′ = 0. The timer output at pin 3 will be HIGH, transistor Q_1 is OFF, and hence the discharge pin 7 is at the open position.

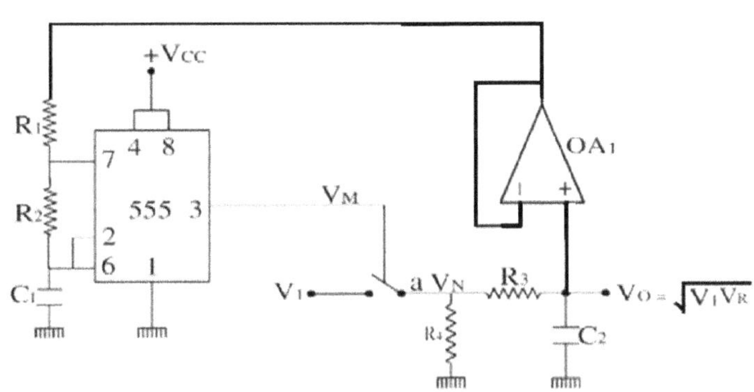

Figure 10.9 Square rooter from 555 astable.

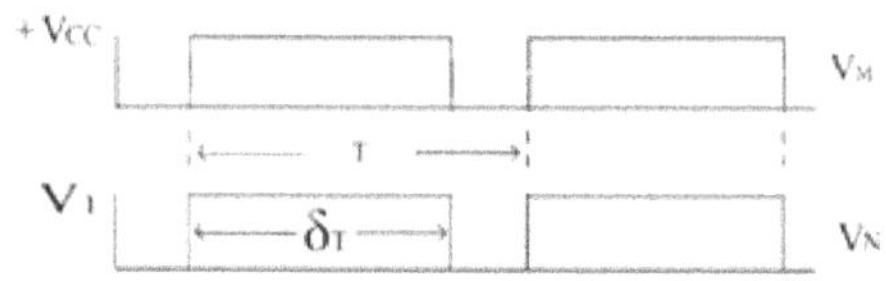

Figure 10.10 Associated waveforms of Figure 10.9.

The capacitor C_1 is charging toward V_O through the resistors R_1 and R_2 with a time constant of $(R_1+R_2)C_1$, and its voltage is rising exponentially. When the capacitor voltage rises above the voltage 2/3 V_{CC}, the output of the upper comparator CMP_1 becomes HIGH, i.e., R = 1, and the output of the lower comparator CMP_2 becomes LOW, i.e., S = 0. The flip flop outputs are Q = 0 and Q′ = 1. The timer output at pin 3 will be LOW, transistor Q_1 is ON, and hence the discharge pin 7 is at GND potential. Now the capacitor C_1 is discharging to GND potential through the resistor R_2 with a time constant of R_2C. When the capacitor voltage falls below 1/3 V_{CC}, the output of the upper comparator CMP_1 becomes LOW, i.e., R = 0, and the output of the lower comparator CMP_2 becomes HIGH, i.e., S = 1. The flip flop outputs are Q = 1 and Q′ = 0. The timer output at pin 3 will be HIGH, transistor Q_1 is OFF, and hence the discharge pin 7 is at the open position.

Now the capacitor starts charging toward V_O, and the cycle therefore repeats to produce periodic pulses at the output pin 3 of the 555 timer.

The ON time of the 555 timer output V_M is inversely proportional to V_O. During the ON time δ_T, the input voltage V_1 is connected to the R_3C_2 low pass filter (switch S_1 is closed). During the OFF time of V_M, zero voltage exists on the R_3C_2 low pass filter (switch S1 is opened). Another rectangular waveform V_N with V_1 as the peak value is generated at the output of switch S_1. The ON time δ_T of this rectangular pulse V_N is given as

$$\delta_T = \frac{V_R}{V_O}T \tag{10.27}$$

where V_R is a constant value.

The R_3C_2 low pass filter gives the average value of this pulse train V_N and is given as

$$V_O = \frac{1}{T}\int_0^{\delta_T} V_1 dt = \frac{V_1}{T}\delta_T$$

$$V_O = \sqrt{V_1 V_R} \tag{10.28}$$

10.5.2 Square Rooter from 555 Astable Multivibrator—Type II

The circuit diagram of a square rooter using the 555 timer astable multivibrator is shown in Figure 10.11, and its associated waveforms are shown in Figure 10.12. Refer to the internal diagram of the 555 timer IC shown in Figure 0.1. Initially when we switch on the power supply, the output of the upper comparator CMP_1 will be LOW, i.e., R = 0, and the output of the lower comparator CMP_2 will be HIGH, i.e., S = 1. The flip flop outputs are Q = 1 and Q′ = 0. The timer output at pin 3 will be HIGH, transistor Q_1 is OFF, and hence the discharge pin 7 is at the open position.

The capacitor C_1 is charging toward $+V_{CC}$ through the resistors R_1 and R_2 with a time constant of $(R_1+R_2)C_1$, and its voltage is rising exponentially. When the capacitor voltage rises above the voltage V_O, the output of the upper comparator CMP_1 becomes HIGH, i.e., R = 1, and the output of the lower comparator CMP_2 becomes LOW, i.e., S = 0. The flip flop outputs are Q = 0 and Q′ = 1. The timer output at pin 3 will be LOW, transistor Q_1 is ON, and hence the discharge pin 7 is at GND potential. Now the capacitor C_1 is discharging to GND potential through the resistor R_2 with a time constant of R_2C. When the capacitor voltage falls below 1/3 V_{CC}, the output of the upper comparator CMP_1 becomes LOW, i.e., R = 0, and the output of the lower comparator CMP_2 becomes HIGH, i.e., S = 1. The flip flop outputs are Q = 1 and Q′ = 0. The timer output at pin 3 will be HIGH, transistor Q_1 is OFF, and hence the discharge pin 7 is at the open position.

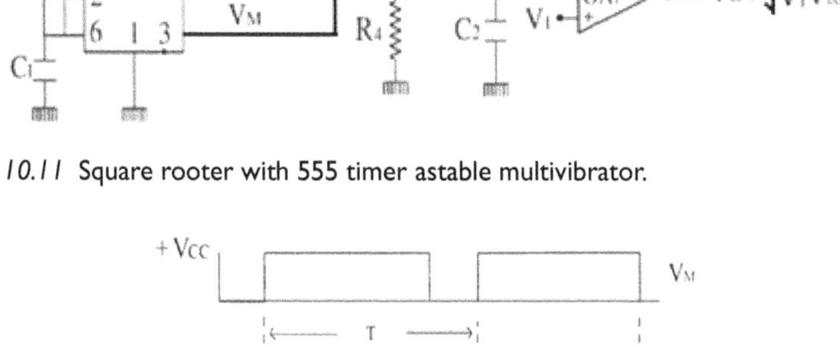

Figure 10.11 Square rooter with 555 timer astable multivibrator.

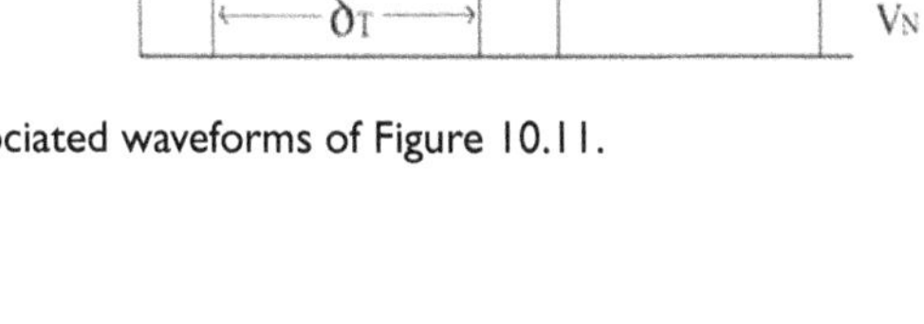

Figure 10.12 Associated waveforms of Figure 10.11.

Now the capacitor starts charging toward $+V_{CC}$, and the cycle therefore repeats to produce periodic pulses at the output pin 3 of the 555 timer.

The ON time of the 555 timer output V_M is proportional to V_O, which is applied at its pin 5. During the ON time δ_T, the voltage V_O is connected to the R_3C_2 low pass filter (switch S_1 is closed). During the OFF time of V_M, zero voltage exists on the R_3C_2 low pass filter (switch S_1 is opened). Another rectangular waveform V_N with V_O as the peak value is generated at the output of switch S_1.

$$\delta_T = \frac{V_O}{V_R} T \tag{10.29}$$

The R_3C_2 low pass filter gives the average value of this pulse train V_N and is given as

$$V_X = \frac{1}{T}\int_0^{\delta_T} V_O dt = \frac{V_O}{T}\delta_T$$

$$V_X = \frac{{V_O}^2}{V_R} \tag{10.30}$$

where V_R is a constant value.

The op-amp OA_1 is kept in a negative closed loop configuration, and a positive dc voltage is ensured in the feedback. Hence its inverting terminal voltage will be equal to its non-inverting terminal voltage, i.e.,

$$V_X = V_1 \tag{10.31}$$

From equations (10.30) and (10.31),

$$V_O = \sqrt{V_1 V_R} \tag{10.32}$$

10.6 SQUARE ROOTER FROM 555 MONOSTABLE MULTIVIBRATOR

10.6.1 Type I

The circuit diagram of a square rooter using the 555 monostable multivibrator is shown in Figure 10.13, and its associated waveforms are shown in Figure 10.14. Refer to the internal diagram of the 555 timer IC shown in Figure 0.1. Initially when the power supply is switched on, the output of the upper comparator CMP_1 will be LOW, i.e., $R = 0$, and the output of the lower comparator CMP_2 will be HIGH, i.e., $S = 1$. The flip flop outputs are $Q = 1$ and $Q' = 0$. The timer output at pin 3 will be HIGH, transistor Q_1 is

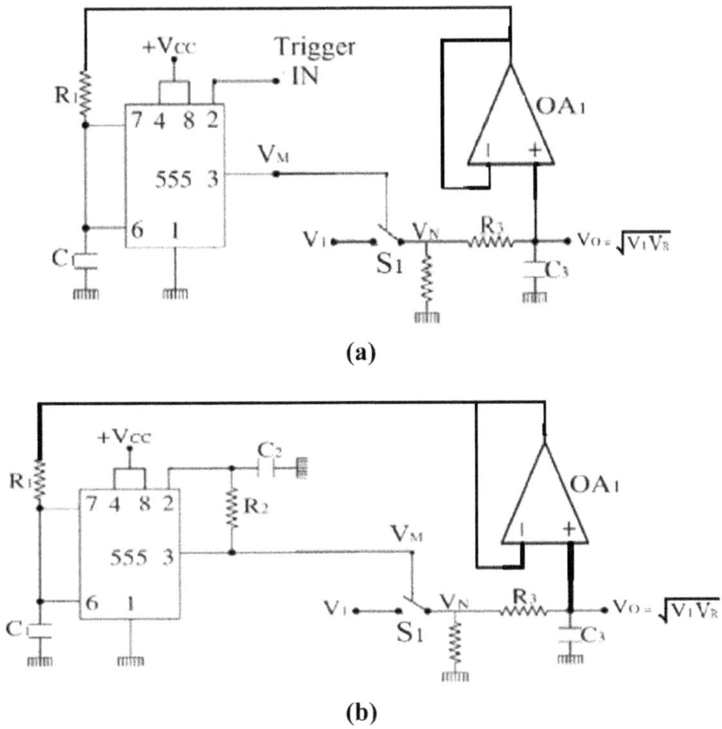

Figure 10.13 (a) Square rooter using 555 timer monostable multivibrator. (b) Square rooter using re-trigger monostable multivibrator.

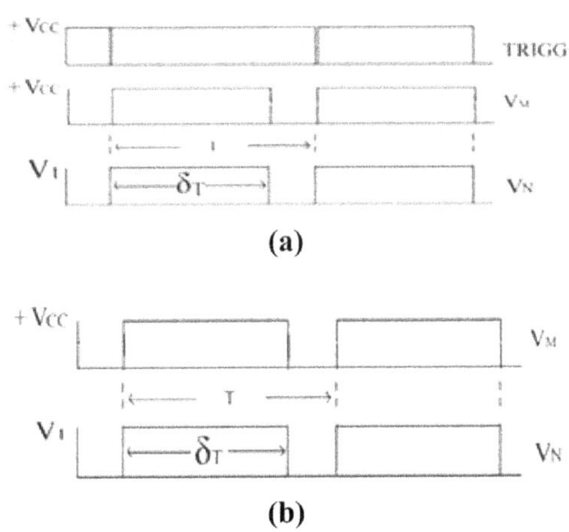

Figure 10.14 (a) Associated waveforms of Figure 10.13(a). (b) Associated waveforms of Figure 10.13(b).

OFF, and hence the discharge pin 7 is at the open position. The capacitor C_1 is charging toward V_O through the resistor R_1. The capacitor voltage is rising exponentially, and when it reaches the value of 2/3 V_{CC}, the output of the upper comparator CMP_1 becomes HIGH, i.e., R = 1, and the output of the lower comparator CMP_2 becomes LOW, i.e., S = 0. The flip flop outputs are Q = 0 and Q′ = 1. The timer output at pin 3 will be LOW, transistor Q_1 is ON, and hence the discharge pin 7 is at GND potential. Now the capacitor C_1 is short circuited, zero voltage exists at pin 6, the output of the upper comparator CMP_1 becomes LOW, i.e., R = 0. A trigger pulse is applied at pin 2, and when the trigger voltage comes down to 1/3 V_{CC}, the output of the lower comparator CMP_2 becomes HIGH, i.e., S = 1. The flip flop outputs are Q = 1 and Q′ = 0. The timer output at pin 3 will be HIGH, transistor Q_1 is OFF, and hence the discharge pin 7 is at the open position.

Now the capacitor C_1 is charging toward V_O, and the sequence therefore repeats for every trigger input pulse.

The ON time of the 555 timer output V_M is inversely proportional to V_O. The output of the 555 timer controls switch S_1. During the ON time δ_T, the input voltage V_1 is connected to the R_3C_3 low pass filter (switch S1 is closed). During the OFF time of V_M, zero voltage exists on the R_3C_3 low pass filter (switch S_1 is opened). Another rectangular waveform V_N, with V_1 as the peak value, is generated at the output of switch S_1. The ON time of rectangular wave is given as

$$\delta_T = \frac{V_R}{V_O} T \tag{10.33}$$

The R_3C_3 low pass filter gives the average value of this pulse train V_N and is given as

$$V_O = \frac{1}{T}\int_0^{\delta_T} V_1 dt = \frac{V_1}{T}\delta_T$$

$$V_O = \sqrt{V_1 V_R} \tag{10.34}$$

where V_R is a constant value.

Figure 10.13(b) shows a re-trigger monostable multivibrator used as an analog square rooter.

10.6.2 Square Rooter from 555 Monostable Multivibrator—Type II

The circuit diagrams of a square rooter using the 555 timer monostable multivibrator are shown in Figure 10.15, and their associated waveforms are shown in Figure 10.16. Refer to the internal diagram of the 555 timer

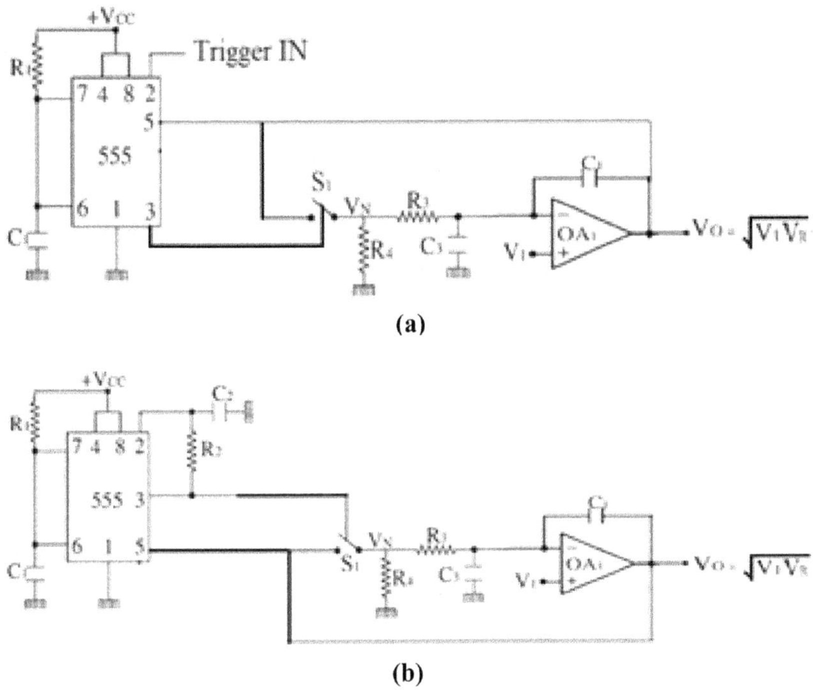

Figure 10.15 (a) Square rooter from 555 monostable. (b) Square rooter with 555 re-trigger monostable multivibrator.

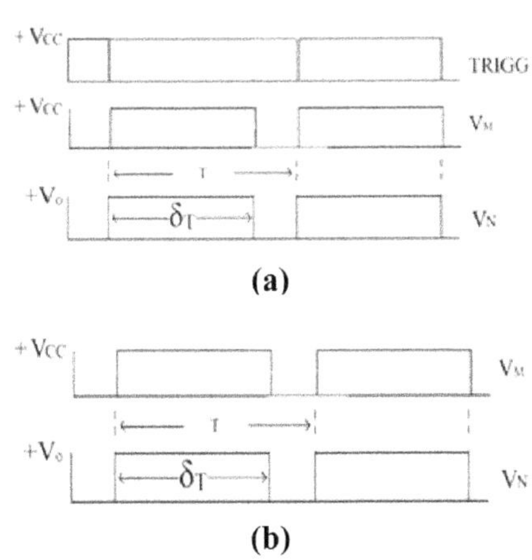

Figure 10.16 (a) Associated waveforms of Figure 10.15(a). (b) Associated waveforms of Figure 10.15(b).

IC shown in Figure 0.1. Initially when the power supply is switched on, the output of the upper comparator CMP_1 will be LOW, i.e., R = 0, and the output of the lower comparator CMP_2 will be HIGH, i.e., S = 1. The flip flop outputs are Q = 1 and Q′ = 0. The timer output at pin 3 will be HIGH, transistor Q_1 is OFF, and hence the discharge pin 7 is at the open position. The capacitor C_1 is charging toward $+V_{CC}$ through the resistor R_1. The capacitor voltage is rising exponentially, and when it reaches the value of V_O, the output of the upper comparator CMP_1 becomes HIGH, i.e., R = 1, and the output of the lower comparator CMP_2 becomes LOW, i.e., S = 0. The flip flop outputs are Q = 0 and Q′ = 1. The timer output at pin 3 will be LOW, transistor Q_1 is ON, and hence the discharge pin 7 is at GND potential. Now the capacitor C_1 is short circuited, zero voltage exists at pin 6, the output of the upper comparator CMP_1 becomes LOW, i.e., R = 0. A trigger pulse is applied at pin 2, and when the trigger voltage comes down to 1/3 V_{CC}, the output of the lower comparator CMP_2 becomes HIGH, i.e., S = 1. The flip flop outputs are Q = 1 and Q′ = 0. The timer output at pin 3 will be HIGH, transistor Q_1 is OFF, and hence the discharge pin 7 is at the open position.

Now the capacitor C_1 is charging toward $+V_{CC}$, and the sequence therefore repeats for every trigger input pulse.

The ON time of the 555 timer output V_M is proportional to V_O, which is applied at its pin 5. The 555 timer output controls the switch S_1. During the ON time δ_T, the output voltage V_O is connected to the R_3C_3 low pass filter (switch S_1 is closed). During the OFF time of V_M, zero voltage exists on the R_3C_2 low pass filter (switch S_1 is opened). Another rectangular waveform V_N with V_O as the peak value is generated at the output of switch S_1. The ON time δ_T of this rectangular waveform V_N is given as

$$\delta_T = \frac{V_O}{V_R} T \tag{10.35}$$

The R_3C_3 low pass filter gives the average value of this pulse train V_N and is given as

$$V_X = \frac{1}{T}\int_0^{\delta_T} V_O dt = \frac{V_O}{T}\delta_T$$

$$V_X = \frac{V_O^{\,2}}{V_R} \tag{10.36}$$

where V_R is a constant value.

The op-amp OA_1 is kept in a negative closed loop configuration, and a positive dc voltage is ensured in the feedback. Hence its inverting terminal voltage will be equal to its non-inverting terminal voltage, i.e.,

$$V_X = V_1 \tag{10.37}$$

From equations (10.36) and (10.37),

$$V_O = \sqrt{V_1 V_R} \tag{10.38}$$

Figure 10.15(b) shows the square rooter with 555 re-trigger monostable multivibrator.

Chapter 11

Multiplexing Time Division Vector Magnitude Circuits—Part I

11.1 SAW TOOTH WAVE REFERENCED VMCs

The circuit diagram of a double multiplexing-averaging time division VMC is shown in Figure 11.1, and its associated waveforms are shown in Figure 11.2. A saw tooth wave V_{S1} of peak value V_R and time period T is generated by the 555 timer.

The comparator OA_3 compares the saw tooth wave V_{S1} with the voltage V_Y and produces a rectangular waveform V_K. The ON time δ_T of V_K is given as

$$\delta_T = \frac{V_Y}{V_R} T \tag{11.1}$$

The rectangular pulse V_K controls the second multiplexer M_2. When V_K is HIGH, the first input voltage V_1 is connected to the R_4C_3 low pass filter ('by' is connected to 'b'). When V_K is LOW, zero voltage is connected to the R_4C_3 low pass filter ('bx' is connected to 'b'). Another rectangular pulse V_M with a maximum value of V_B is generated at the multiplexer M_2 output. The R_4C_3 low pass filter gives the average value of this pulse train V_M and is given as

$$V_X = \frac{1}{T}\int_0^{\delta_T} V_B dt = \frac{V_B}{T}\delta_T \tag{11.2}$$

$$V_X = \frac{V_B V_Y}{V_R} \tag{11.3}$$

The op-amp OA_4 is configured in a negative closed loop feedback, and a positive dc voltage is ensured in the feedback loop. Hence its inverting terminal voltage must be equal to its non-inverting terminal voltage.

$$V_1 = V_X \tag{11.4}$$

DOI: 10.1201/9781003362968-11

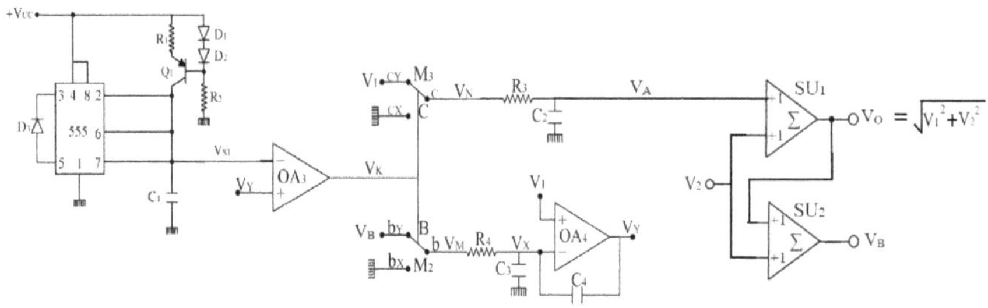

Figure 11.1 Double multiplexing-averaging VMC.

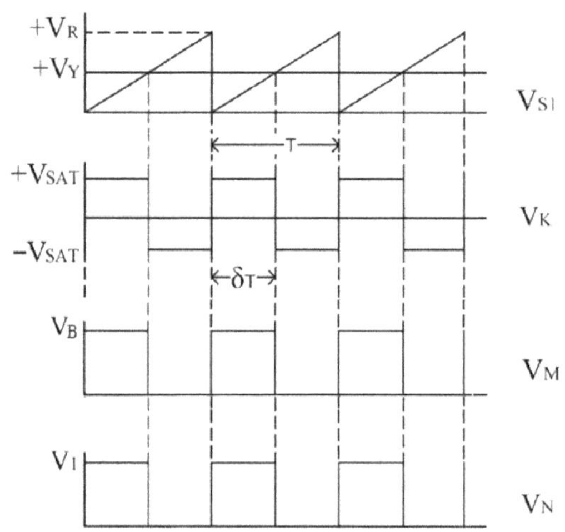

Figure 11.2 Associated waveforms of Figure 11.1.

From equations (11.3) and (11.4),

$$V_Y = \frac{V_1 V_R}{V_B} \tag{11.5}$$

The rectangular pulse V_K also controls the third multiplexer M_3. When V_K is HIGH, the input voltage V_1 is connected to the R_3C_2 low pass filter ('cy' is connected to 'c'). When V_K is LOW, zero voltage is connected to the R_3C_2 low pass filter ('cx' is connected to 'c'). Another rectangular pulse V_N with a maximum value of V_1 is generated at the multiplexer M_3 output. The R_3C_2 low pass filter gives the average value of this pulse train V_N and is given as

$$V_A = \frac{1}{T}\int_0^{\delta_T} V_1 dt = \frac{V_1}{T}\delta_T \tag{11.6}$$

Equations (11.1) and (11.5) in (11.6) give

$$V_A = \frac{V_1^2}{V_B} \tag{11.7}$$

The output of the summer SU_2 will be

$$V_B = V_O + V_2 \tag{11.8}$$

The output of the summer SU_1 will be

$$V_O = V_A + V_2 \tag{11.9}$$

Equations (11.7) and (11.8) in (11.9) give

$$V_O = \frac{V_1^2}{V_O + V_2} + V_2 \tag{11.10}$$

$$V_O^2 = V_1^2 + V_2^2 \tag{11.11}$$

$$V_O = \sqrt{V_1^2 + V_2^2} \tag{11.12}$$

11.2 TRIANGULAR WAVE BASED TIME DIVISION VMCs

The circuit diagram of a triangular wave referenced time division VMC is shown in Figure 11.3, and its associated waveforms are shown in Figure 11.4. A triangular wave V_{T1} of $\pm V_T$ peak to peak values and time period T is generated by the 555 timer. The comparator OA_3 compares the triangular wave V_{T1} with the voltage V_Y and produce an asymmetrical rectangular wave V_K. From Figure 11.4, it is observed that

$$T_1 = \frac{V_T - V_Y}{2V_T} T T_2 = \frac{V_T + V_Y}{2V_T} T T = T_1 + T_2 \tag{11.13}$$

The rectangular wave V_K controls the multiplexer M_1, which connects $+V_B$ to its output duringT_2 ('ay' is connected to 'a') and $-V_B$ to its output during T_1 ('ax' is connected to 'a'). Another asymmetrical rectangular waveform V_N is generated at the multiplexer M_1 output with $\pm V_B$ peak to peak values. The R_4C_2 low pass filter gives the the average value of V_N and is given as

$$V_X = \frac{1}{T}\left[\int_O^{T_2} V_B\,dt + \int_{T_2}^{T_1+T_2} (-V_B)\,dt\right] = \frac{V_B}{T}[T_2 - T_1]$$
$$V_X = \frac{V_B V_Y}{V_T} \tag{11.14}$$

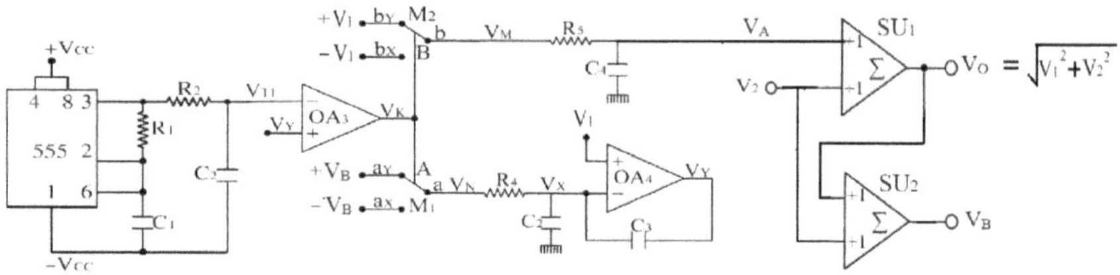

Figure 11.3 Triangular wave based time division VMC.

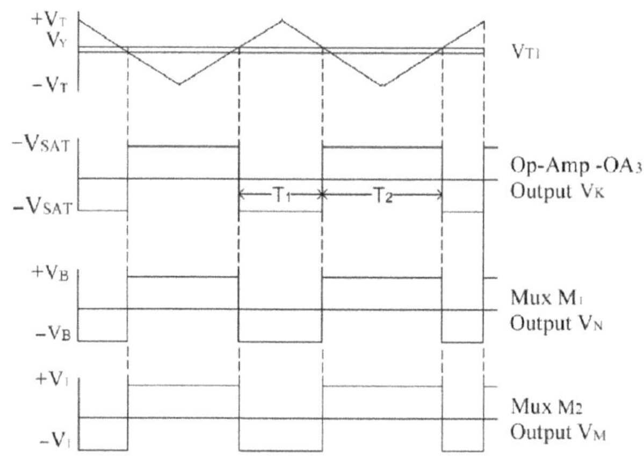

Figure 11.4 Associated waveforms of Figure 11.3.

The op-amp OA_4 is configured in a negative closed loop feedback, and a positive dc voltage is ensured in the feedback loop. Hence its inverting terminal voltage must be equal to its non-inverting terminal voltage, i.e.,

$$V_X = V_1 \tag{11.15}$$

From equations (11.14) and (11.15),

$$V_Y = \frac{V_1 V_T}{V_B} \tag{11.16}$$

The rectangular wave V_K also controls the multiplexer M_2, which connects $+V_1$ to its output during T_2 ('by' is connected to 'b') and connects $-V_1$ to output during T_1 ('bx' is connected to 'b'). Another asymmetrical rectangular wave V_M is generated at the multiplexer M_2 output with $\pm V_1$ peak to peak values. The R_5C_4 low pass filter gives the average value V_A and is given as

$$V_A = \frac{1}{T}\left[\int_{O}^{T_2} V_1\,dt + \int_{T_2}^{T_1+T_2} (-V_1)\,dt\right] = \frac{V_1}{T}[T_2 - T_1] \tag{11.17}$$

$$V_A = \frac{V_1 V_Y}{V_T}$$

Equation (11.16) in (11.17) gives

$$V_A = \frac{V_1^{\,2}}{V_B} \tag{11.18}$$

The output of the summer SU_2 will be

$$V_B = V_O + V_2 \tag{11.19}$$

The output of the summer SU_1 will be

$$V_O = V_A + V_2 \tag{11.20}$$

Equations (11.18) and (11.19) in (11.20) give

$$V_O = \frac{V_1^{\,2}}{V_O + V_2} + V_2 \tag{11.21}$$

$$V_O^{\,2} = V_1^{\,2} + V_2^{\,2}$$

$$V_O = \sqrt{V_1^{\,2} + V_2^{\,2}} \tag{11.22}$$

11.3 VMC FROM 555 ASTABLE MULTIVIBRATOR

The circuit diagram of VMC using the 555 astable is shown in Figure 11.5, and its associated waveforms are shown in Figure 11.6. Refer to the internal diagram of the 555 timer IC shown in Figure 0.1. Initially when we switch on the power supply, the output of the upper comparator CMP_1 will be LOW, i.e., R = 0, and the output of the lower comparator CMP_2 will be HIGH, i.e., S = 1. The flip flop outputs are Q = 1 and Q′ = 0. The timer output at pin 3 will be HIGH, transistor Q_1 is OFF, and hence the discharge pin 7 is at the open position.

The capacitor C_1 is charging toward V_B through the resistors R_1 and R_2 with a time constant of $(R_1+R_2)C_1$, and its voltage is rising exponentially. When the capacitor voltage is rising above the voltage V_1, the output of the upper comparator CMP_1 becomes HIGH, i.e., R = 1, and the output of the lower comparator CMP_2 becomes LOW, i.e., S = 0. The flip flop outputs are

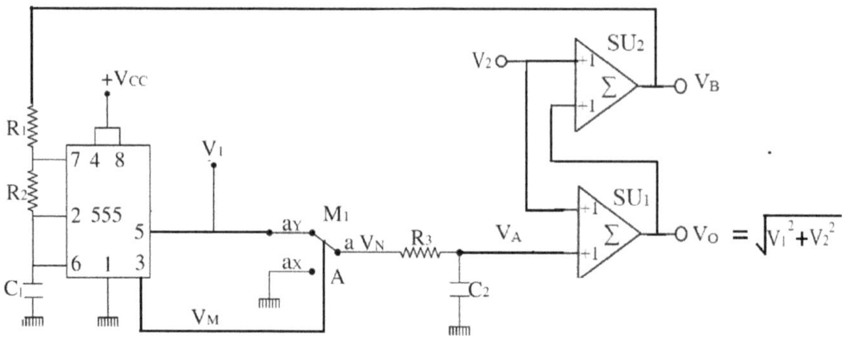

Figure 11.5 555 timer astable as VMC.

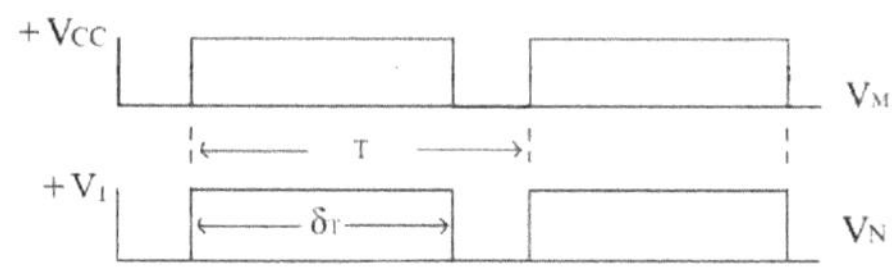

Figure 11.6 Associated waveforms of Figure 11.5.

Q = 0 and Q′ = 1. The timer output at pin 3 will be LOW, transistor Q_1 is ON, and hence the discharge pin 7 is at GND potential. Now the capacitor C_1 is discharging to GND potential through the resistor R_2 with a time constant of R_2C. When the capacitor voltage falls below 1/3 V_{CC}, the output of the upper comparator CMP_1 becomes LOW, i.e., R = 0, and the output of the lower comparator CMP_2 becomes HIGH, i.e., S = 1. The flip flop outputs are Q = 1 and Q′ = 0. The timer output at pin 3 will be HIGH, transistor Q_1 is OFF, and hence the discharge pin 7 is at the open position.

Now the capacitor starts charging toward V_B, and the cycle therefore repeats to produce periodic pulses at the output pin 3 of the 555 timer.

The ON time δ_T of the 555 timer output V_M is (1) proportional to V_1, which is applied at its pin 5 and (2) inversely proportional to the voltage V_B. During the ON time δ_T, V_1 is connected to V_N ('ay' is connected to 'a'). During the OFF time of the waveform V_M, zero voltage is connected to V_N ('ax' is connected to 'a'). Another rectangular waveform V_N with V_1 as the peak value is generated at the output of the multiplexer M_1.

$$\delta_T = K\frac{V_1}{V_B}T \tag{11.23}$$

The R_3C_2 low pass filter gives the average value of this pulse train V_N and is given as

$$V_A = \frac{1}{T}\int_0^{\delta_T} V_1 dt = \frac{V_1}{T}\delta_T$$

$$V_A = \frac{V_1^2}{V_B}K \tag{11.24}$$

where K is a constant value.

The output of the summer SU_2 will be

$$V_B = V_O + V_2 \tag{11.25}$$

The output of the summer SU_1 will be

$$V_O = V_A + V_2 \tag{11.26}$$

Equations (11.24) and (11.25) in (11.26) give

$$V_O = \frac{V_1^2}{V_O + V_2} + V_2 \tag{11.27}$$

$$V_O^2 = V_1^2 + V_2^2$$

$$V_O = \sqrt{V_1^2 + V_2^2} \tag{11.28}$$

11.4 SQUARE WAVE REFERENCED VMC

The circuit diagram of a square wave referenced VMC is shown in Figure 11.7, and its associated waveform is shown in Figure 11.8. A square waveform V_C is generated by the 555 timer. During the LOW of the square wave, the multiplexer M_1 connects 'ax' to 'a', an integrator formed by resistor R_1, capacitor C_1, and op-amp OA_1, integrates the first input voltage $-V_B$. The integrated output will be

$$V_{S1} = -\frac{1}{R_1C_1}\int -V_B dt = \frac{V_B}{R_1C_1}t \tag{11.29}$$

A positive going ramp Vs_1 is generated at the output of the op-amp OA_1. During the HIGH of the square waveform, the multiplexer M_1 connects 'ay' to 'a', and hence the capacitor C_1 is shorted so that op-amp OA_1 output becomes zero. The cycle therefore repeats to provide a semi-saw tooth wave of peak value V_R at the output of the op-amp $OA_{1.}$

From the waveforms shown in Figure 11.8 and equation (11.29), at $t = T/2$, $V_{S1} = V_{R.}$

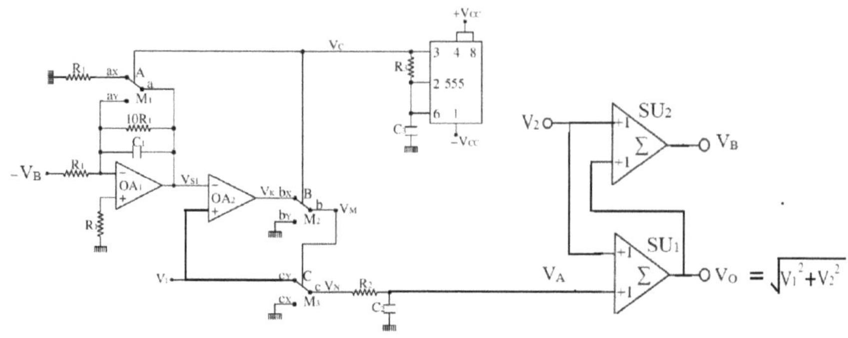

Figure 11.7 Square wave referenced VMC.

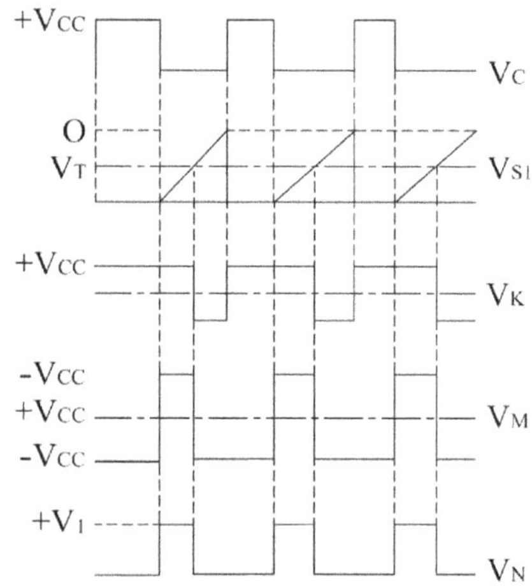

Figure 11.8 Associated waveforms of Figure 11.7.

$$V_R = \frac{V_B}{R_1C_1}\frac{T}{2}$$

$$T/2 = \frac{V_R}{V_B}R_1C_1 \tag{11.30}$$

The comparator OA_2 compares the semi-saw tooth wave V_{S1} of peak value V_R with the input voltage V_1 and produces a rectangular waveform V_K at its output. The square wave V_C controls the second multiplexer M_2. The multiplexer M_2 connects zero volts during the HIGH of V_C and V_K during the LOW of V_C. Another rectangular waveform V_M is generated at the output of the multiplexer M_2. The ON time δ_T of this rectangular waveform V_M is given as

$$\delta_T = \frac{V_1}{V_R}\frac{T}{2} \tag{11.31}$$

The rectangular pulse V_M controls the third multiplexer M_3. When the V_M is HIGH, a third input voltage V_3 is connected to the R_2C_2 low pass filter ('cy' is connected to 'c'). When the V_M is LOW, zero voltage is connected to the R_2C_2 low pass filter ('cx' is connected to 'c'). Another rectangular pulse V_N with a maximum value of V_1 is generated at the multiplexer M_2 output. The R_2C_2 low pass filter gives the average value of this pulse train V_N and is given as

$$V_A = \frac{1}{T}\int_0^{\delta_T} V_1 dt$$

$$V_A = \frac{V_1}{T}\delta_T \tag{11.32}$$

From equations (11.30)–(11.32),

$$V_A = \frac{V_1^2}{V_B}\frac{R_1C_1}{T} \tag{11.33}$$

Let $T = R_1C_1$.

$$V_A = \frac{V_1^2}{V_B} \tag{11.34}$$

The output of the summer SU_2 will be

$$V_B = V_O + V_2 \tag{11.35}$$

The output of the summer SU_1 will be

$$V_O = V_A + V_2 \tag{11.36}$$

Equations (11.34) and (11.35) in (11.36) give

$$V_O = \frac{V_1^2}{V_O + V_2} + V_2 \tag{11.37}$$

$$V_O^2 = V_1^2 + V_2^2$$

$$V_O = \sqrt{V_1^2 + V_2^2} \tag{11.38}$$

11.5 VMC FROM 555 MONOSTABLE MULTIVIBRATOR

The circuit diagrams of a VMC using the 555 monostable are shown in Figure 11.9, and their associated waveforms are shown in Figure 11.10. Refer to the internal diagram of the 555 timer IC shown in Figure 0.1. Initially when the power supply is switched on, the output of the upper comparator CMP_1 will be LOW, i.e., R = 0, and the output of the lower comparator CMP_2 will be HIGH, i.e., S = 1. The flip flop outputs are Q = 1 and Q′ = 0. The timer output at pin 3 will be HIGH, transistor Q_1 is OFF, and hence the discharge pin 7 is at the open position. The capacitor C_1 is charging toward V_B through the resistor R_1. The capacitor voltage is rising exponentially and when it reaches the value of V_1, the output of the upper comparator CMP_1

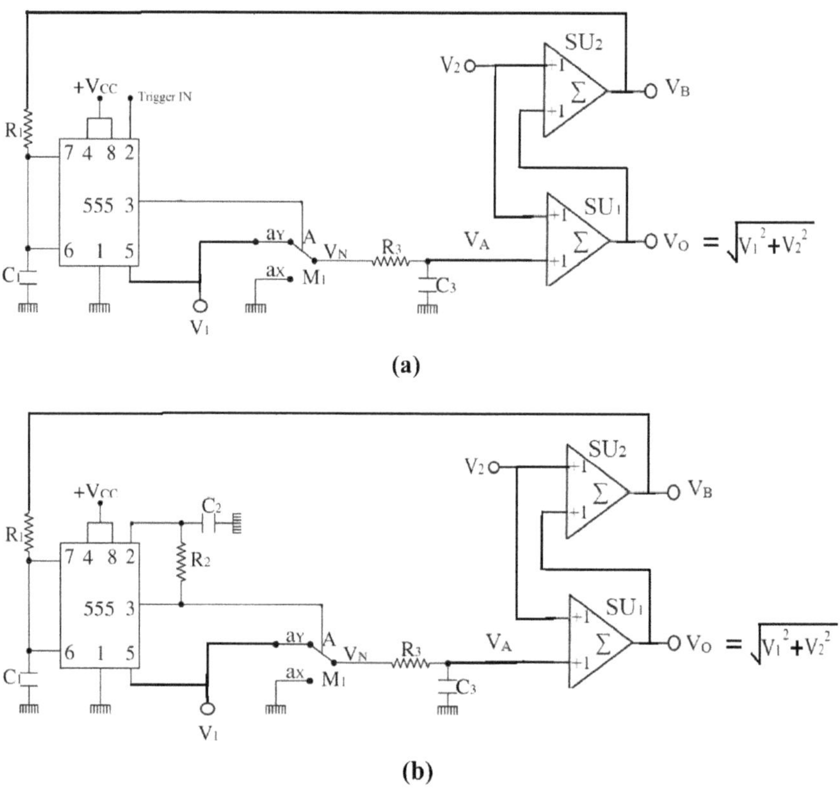

Figure 11.9 (a) 555 monostable as VMC. (b) 555 re-trigger monostable as VMC.

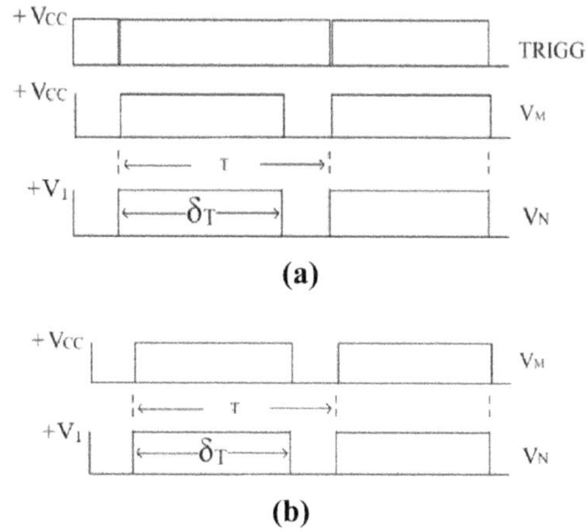

Figure 11.10 (a) Associated waveforms of Figure 11.9(a). (b) A ssociated waveforms of Figure 11.9(b).

becomes HIGH, i.e., R = 1, and the output of the lower comparator CMP_2 becomes LOW, i.e., S = 0. The flip flop outputs are Q = 0 and Q′ = 1. The timer output at pin 3 will be LOW, transistor Q_1 is ON, and hence the discharge pin 7 is at GND potential. Now the capacitor C_1 is short circuited, zero voltage exists at pin 6, and the output of the upper comparator CMP_1 becomes LOW, i.e., R = 0. A trigger pulse is applied at pin 2, and when the trigger voltage comes down to 1/3 V_{CC}, the output of the lower comparator CMP_2 becomes HIGH, i.e., S = 1. The flip flop outputs are Q = 1 and Q′ = 0. The timer output at pin 3 will be HIGH, transistor Q_1 is OFF, and hence the discharge pin 7 is at the open position.

Now the capacitor C_1 is charging toward V_B, and the sequence therefore repeats for every trigger input pulse.

The ON time of the 555 timer output V_M is (1) proportional to V_1, which is applied at its pin 5 and (2) inversely proportional to the voltage V_B. During the ON time δ_T, V_1 is connected to the R_3C_3 low pass filter ('ay' is connected to 'a'). During the OFF time of the waveform V_M, zero voltage is connected to the R_3C_3 low pass filter. Another rectangular waveform V_N with V_1 as the peak value is generated at the output of multiplexer M_1. The ON time δ_T of this rectangular waveform V_N is given as

$$\delta_T = K\frac{V_1}{V_B}T \tag{11.39}$$

The R_3C_3 low pass filter gives the average value of this pulse train V_N and is given as

$$V_A = \frac{1}{T}\int_0^{\delta_T} V_1 dt = \frac{V_1}{T}\delta_T$$

$$V_A = \frac{V_1^2}{V_B}K \tag{11.40}$$

where K is a constant value.

The output of the summer SU_2 will be

$$V_B = V_O + V_2 \tag{11.41}$$

The output of the summer SU_1 will be

$$V_O = V_A + V_2 \tag{11.42}$$

Equations (11.40) and (11.41) in (11.42) give

$$V_O = \frac{V_1^2}{V_O + V_2} + V_2 \tag{11.43}$$

$$V_O^2 = V_1^2 + V_2^2$$

$$V_O = \sqrt{V_1^2 + V_2^2} \tag{11.44}$$

The VMC using an auto-trigger monostable multivibrator is shown in Figure 11.9(b).

11.6 TIME DIVISION VMC WITH NO REFERENCE

A VMC using the time division principle without using any reference clock is shown in Figure 11.11, and its associated waveforms are shown in Figure 11.12.

Initially the 555 timer output is HIGH. The multiplexer M_1 connects $-V_B$ to the differential integrator composed of the resistor R_1, capacitor C_1, and op-amp OA_1 ('ay' is connected to 'a'). The output of differential integrator will be

$$V_{T1} = \frac{1}{R_1C_1}\int (V_1 + V_B)dt$$

$$V_{T1} = \frac{(V_B + V_1)}{R_1C_1}t \tag{11.45}$$

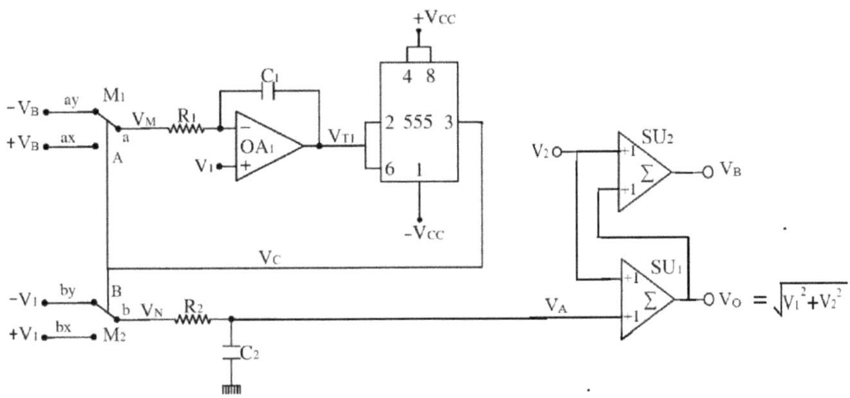

Figure 11.11 Time division VMC without reference clock.

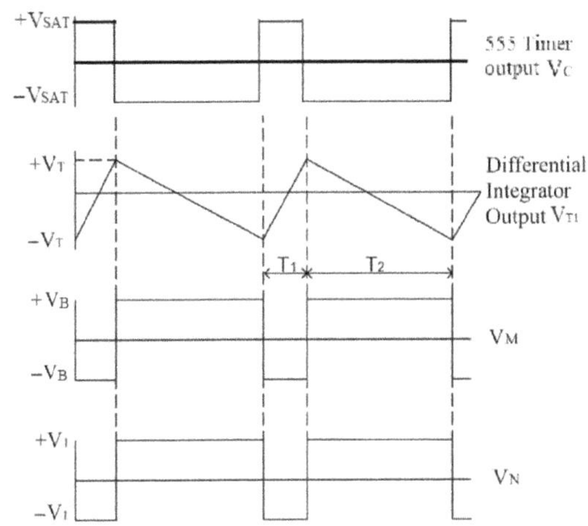

Figure 11.12 Associated waveforms of Figure 11.11.

The output of the differential integrator is rising toward positive saturation, and when it reaches the voltage level of $+V_T$, the 555 timer output becomes LOW. The multiplexer M_1 connects $+V_B$ to the differential integrator composed of the resistor R_1, capacitor C_1, and op-amp OA_1 ('ax' is connected to 'a'). Now the output of differential integrator will be

$$V_{T1} = \frac{1}{R_1C_1}\int (V_1 - V_B)dt$$

$$V_{T1} = -\frac{(V_B - V_1)}{R_1C_1}t \tag{11.46}$$

The output of the differential integrator reverses toward negative saturation, and when it reaches the voltage level $-V_T$, the 555 timer output becomes HIGH, and the cycle therefore repeats, to give an asymmetrical rectangular wave V_C at the output of the 555 timer.

$$V_T = \frac{V_{CC}}{3} \tag{11.47}$$

From the waveforms shown in Figure 11.12, it is observed that

$$T_1 = \frac{V_B - V_1}{2V_B} T, T_2 = \frac{V_B + V_1}{2V_B} T, T = T_1 + T_2 \tag{11.48}$$

The asymmetrical rectangular wave V_C controls another multiplexer M_2. The multiplexer M_2 connects $+V_1$ during the ON time T_2 ('by' is connected to 'b') and $-V_1$ during the OFF time T_1 of the rectangular wave V_C ('bx' is connected to 'b'). Another rectangular wave V_N is generated at the multiplexer M_2 output. The R_2C_2 low pass filter gives the average value of this pulse train V_N and is given as

$$V_A = \frac{1}{T}\left[\int_O^{T_2} V_1\,dt + \int_{T_2}^{T_1+T_2} (-V_1)\,dt\right]$$

$$V_A = \frac{V_1(T_2 - T_1)}{T} \tag{11.49}$$

Equation (11.48) in (11.49) gives

$$V_A = \frac{V_1^2}{V_B} \tag{11.50}$$

The output of the summer SU_2 will be

$$V_B = V_O + V_2 \tag{11.51}$$

The output of the summer SU_1 will be

$$V_O = V_A + V_2 \tag{11.52}$$

Equations (11.50) and (11.51) in (11.52) give

$$V_O = \frac{V_1^2}{V_O + V_2} + V_2 \tag{11.53}$$

$$V_O{}^2 = V_1{}^2 + V_2{}^2$$

$$V_O = \sqrt{V_1{}^2 + V_2{}^2} \qquad (11.54)$$

Chapter 12

Multiplexing Time Division VMC—Part II

12.1 TIME DIVISION VMC WITH NO REFERENCE—TYPE I

A VMC using the time division principle without using any reference clock is shown in Figure 12.1, and its associated waveforms are shown in Figure 12.2.

Initially the 555 timer output is HIGH. The multiplexer M_1 connects $-V_1$ to the inverting terminal of the differential integrator ('ay' is connected 'a'). The output of the differential integrator will be

$$V_{T1} = \frac{1}{R_1C_1}\int (V_A + V_1)dt$$

$$V_{T1} = \frac{(V_A + V_1)}{R_1C_1}t \tag{12.1}$$

The output of the differential integrator rises toward positive saturation, and when it reaches the voltage level of $+V_T$, the 555 timer output becomes LOW. The multiplexer M_1 connects $+V_1$ to the inverting terminal of the differential integrator. Now the output of the differential integrator will be

$$V_{T1} = \frac{1}{R_1C_1}\int (V_A - V_1)dt$$

$$V_{T1} = -\frac{(V_1 - V_A)}{R_1C_1}t \tag{12.2}$$

The output of the differential integrator reverses toward negative saturation, and when it reaches the voltage level $-V_T$, the 555 timer output becomes HIGH, and the cycle therefore repeats, to give an asymmetrical rectangular wave V_C at the output of the 555 timer.

$$V_T = \frac{V_{CC}}{3} \tag{12.3}$$

DOI: 10.1201/9781003362968-12

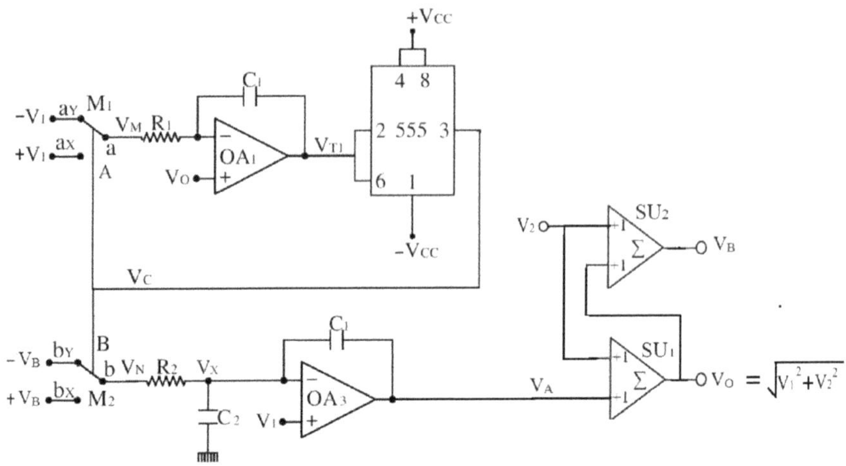

Figure 12.1 VMC without reference clock.

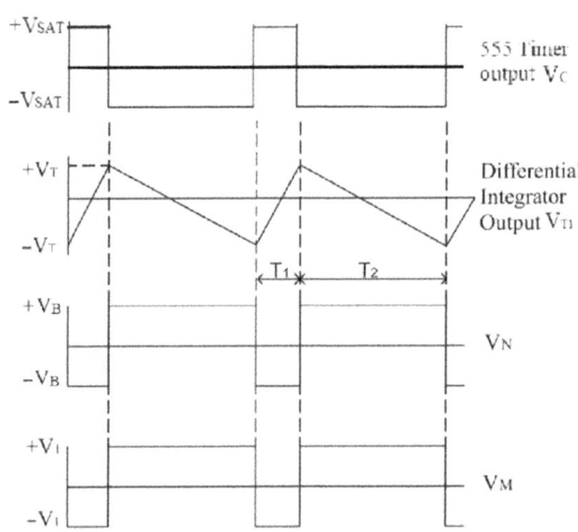

Figure 12.2 Associated waveforms of Figure 12.1.

From the waveforms shown in Figure 12.2, it is observed that

$$T_1 = \frac{V_1 - V_A}{2V_1}T, T_2 = \frac{V_1 + V_A}{2V_1}T, T = T_1 + T_2 \tag{12.4}$$

The asymmetrical rectangular wave V_C controls another multiplexer M_2. The multiplexer M_2 connects $-V_B$ during the ON time T_2 ('by' is connected to 'b') and $+V_B$ during the OFF time T_1 of the rectangular wave V_C ('bx' is connected to 'b'). Another rectangular wave V_N with $\pm V_B$ as the peak to peak value is generated at the multiplexer M_2 output. The R_2C_2 low pass filter gives the average value of this pulse train V_N and is given as

$$V_X = \frac{1}{T}\left[\int_{O}^{T_2} V_B\,dt + \int_{T_2}^{T_1+T_2} (-V_B)\,dt\right]$$

$$V_X = \frac{V_B(T_2 - T_1)}{T} \tag{12.5}$$

Equations (12.4) in (12.5) give

$$V_X = \frac{V_A V_B}{V_1} \tag{12.6}$$

The op-amp OA_3 is at a negative closed loop configuration, and a positive dc voltage is ensured in the feedback loop. Hence its non-inverting terminal voltage is equal to its inverting terminal voltage, i.e.,

$$V_1 = V_X \tag{12.7}$$

From equations (12.6) and (12.7),

$$V_A = \frac{V_1^2}{V_B} \tag{12.8}$$

The output of the summer SU_2 will be

$$V_B = V_O + V_2 \tag{12.9}$$

The output of the summer SU_1 will be

$$V_O = V_A + V_2 \tag{12.10}$$

Equations (12.8) and (12.9) in (12.10) give

$$V_O = \frac{V_1^2}{V_O + V_2} + V_2 \tag{12.11}$$

$$V_O^2 = V_1^2 + V_2^2$$

$$V_O = \sqrt{V_1^2 + V_2^2} \tag{12.12}$$

12.2 TIME DIVISION VMC WITH NO REFERENCE—TYPE II

The VMC using the time division principle without using any reference clock is shown in Figure 12.3, and its associated waveforms are shown in Figure 12.4.

A rectangular pulse V_M with a V_{SAT} peak value is generated at output pin 3 of the 555 timer.

The ON time δ_T of this rectangular waveform V_M is given as

$$\delta_T = K\frac{V_1}{V_B}T \tag{12.13}$$

Another rectangular pulse V_N with a maximum value of V_1 is generated at pin 7 of the 555 timer. The R_2C_2 low pass filter gives the average value of this pulse train V_N and is given as

$$V_A = \frac{1}{T}\int_0^{\delta_T} V_1 dt \tag{12.14}$$

$$V_A = \frac{V_1}{T}\delta_T \tag{12.15}$$

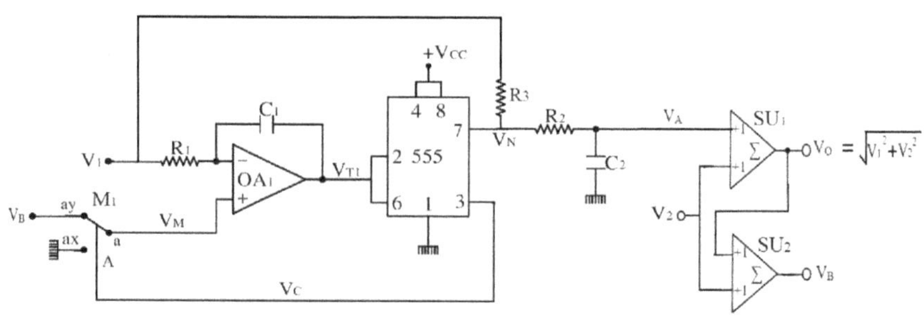

Figure 12.3 Time division VMC without reference clock.

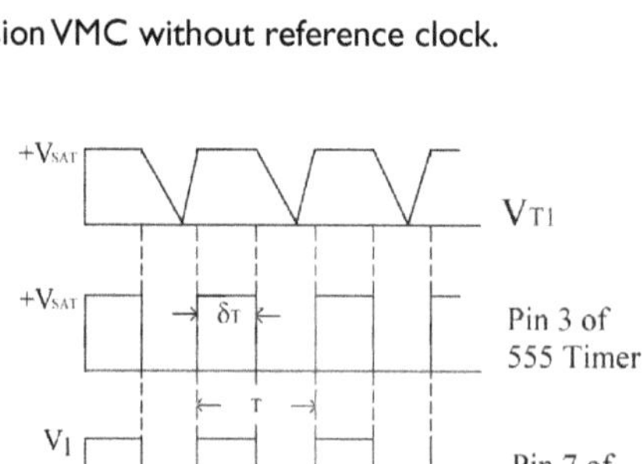

Figure 12.4 Associated waveforms of Figure 12.3.

Equation (12.13) in (12.15) gives

$$V_A = \frac{V_1^{\,2}}{V_B}K \tag{12.16}$$

The output of the summer SU_2 will be

$$V_B = V_O + V_2 \tag{12.17}$$

The output of the summer SU_1 will be

$$V_O = V_A + V_2 \tag{12.18}$$

Equations (12.16) and (12.17) in (12.18) give

$$V_O = \frac{V_1^{\,2}}{V_O + V_2} + V_2 \tag{12.19}$$

$$V_O^{\,2} = V_1^{\,2} + V_2^{\,2}$$

$$V_O = \sqrt{V_1^{\,2} + V_2^{\,2}} \tag{12.20}$$

where K is a constant value.

12.3 TIME DIVISION VMC WITH NO REFERENCE—TYPE III

The VMC using the time division principle without using any reference clock is shown in Figure 12.5, and its associated waveforms are shown in Figure 12.6.

A rectangular pulse V_M with a V_{SAT} peak value is generated at output pin 3 of the 555 timer. The ON time δ_T of this rectangular waveform V_M is given as

$$\delta_T = K\frac{V_A}{V_1}T \tag{12.21}$$

Another rectangular wave V_N with $+V_B$ as the peak to peak value is generated at output pin 7 of the 555 timer. The R_2C_2 low pass filter gives the average value of this pulse train V_N and is given as

$$V_X = \frac{1}{T}\int_0^{\delta_T} V_B dt \tag{12.22}$$

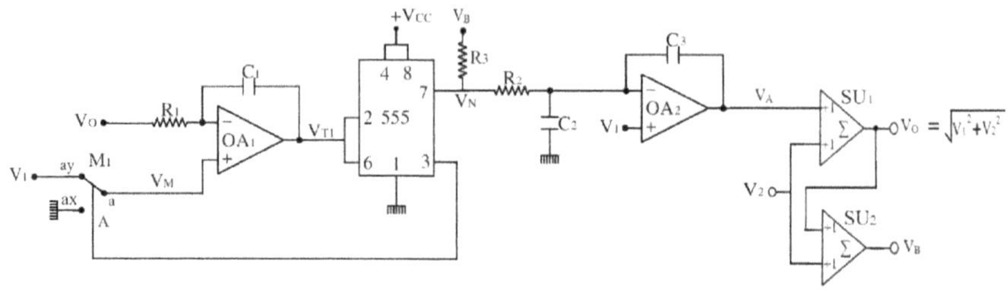

Figure 12.5 VMC without reference clock.

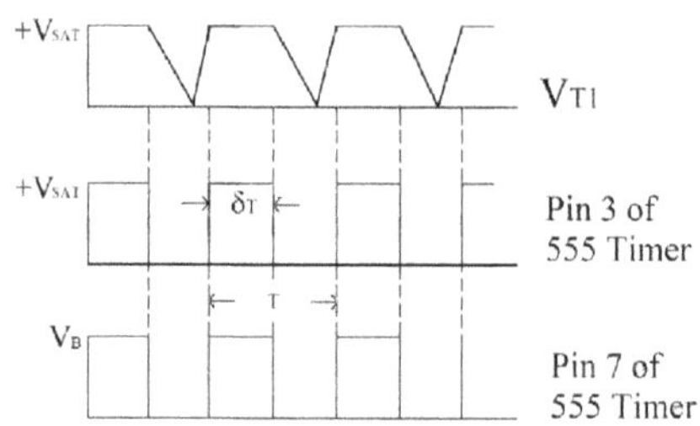

Figure 12.6 Associated waveforms of Figure 12.5.

$$V_X = \frac{V_B}{T}\delta_T \tag{12.23}$$

Equation (12.20) in (12.23) gives

$$V_X = \frac{V_A V_B}{V_1} \tag{12.24}$$

The op-amp OA_2 is at a negative closed loop configuration, and a positive dc voltage is ensured in the feedback loop. Hence its non-inverting terminal voltage is equal to its inverting terminal voltage, i.e.,

$$V_1 = V_X \tag{12.25}$$

From equations (12.24) and (12.25),

$$V_A = \frac{V_1^2}{V_B} \tag{12.26}$$

The output of the summer SU_2 will be

$$V_B = V_O + V_2 \tag{12.27}$$

The output of the summer SU_1 will be

$$V_O = V_A + V_2 \tag{12.28}$$

Equations (12.26) and (12.27) in (12.28) give

$$V_O = \frac{V_1^2}{V_O + V_2} + V_2 \tag{12.29}$$

$$V_O^2 = V_1^2 + V_2^2$$

$$V_O = \sqrt{V_1^2 + V_2^2} \tag{12.30}$$

12.4 TIME DIVISION VMC WITH NO REFERENCE—TYPE IV

The VMC using the time division principle without using any reference clock is shown in Figure 12.7, and its associated waveforms are shown in Figure 12.8.

Initially the 555 timer output pin 3 is HIGH, and pin 7 of the 555 timer is opened. The output of the differential integrator, composed of the resistor R_1, capacitor C_1, and op-amp OA_1, will be

$$V_{T1} = \frac{1}{R_1C_1}\int (V_B - V_1)dt + V_B$$

$$V_{T1} = \frac{(V_B - V_1)}{R_1C_1}t + V_B \tag{12.31}$$

The output of the differential integrator rises toward positive saturation, and when it reaches the voltage level of $2V_{CC}/3$, the 555 timer output pin 3 becomes LOW. Pin 7 of the 555 timer goes to GND. Now the output of the

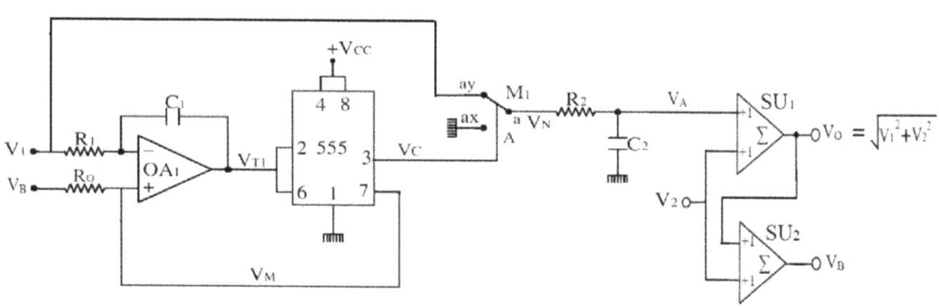

Figure 12.7 Time division VMC without reference clock.

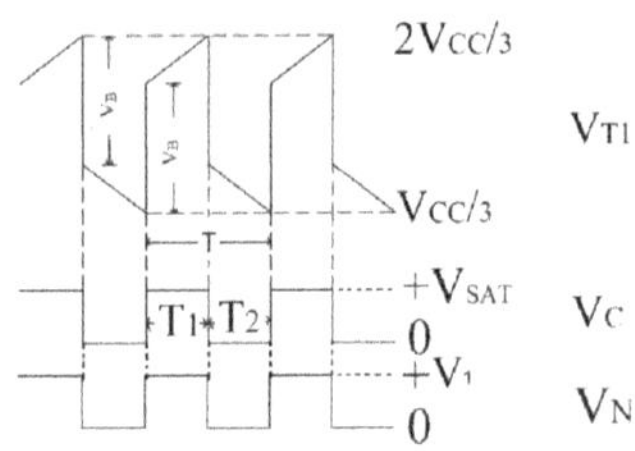

Figure 12.8 Associated waveforms of Figure 12.7.

differential integrator, composed of resistor R_1, capacitor C_1, and op-amp OA_1, will be

$$V_{T1} = \frac{1}{R_1C_1}\int(0 - V_1)dt + V_B$$

$$V_{T1} = -\frac{V_1}{R_1C_1}t + V_B \tag{12.32}$$

The output of the differential integrator reverses toward negative saturation, and when it reaches the voltage level ($V_{CC}/3$), the 555 timer output becomes HIGH, and the cycle therefore repeats, to give (1) an asymmetrical rectangular wave V_C at the output of the 555 timer and (2) another asymmetrical rectangular wave V_N at the output of multiplexer M_1 with a peak value of V_1.

From the waveforms shown in Figure 12.8, it is observed that the differential integrator output has an abrupt fall and rise by V_B due to the fact that the capacitor C cannot change voltage across it abruptly.

From the waveforms shown in Figure 12.8, it is observed that

$$T_1 = \frac{R_1C_1}{V_B - V_1}V \tag{12.33}$$

$$T_2 = \frac{R_1C_1}{V_B}V \tag{12.34}$$

$$T = T_1 + T_2 = \frac{R_1C_1V_BV}{(V_B - V_1)V_1} \tag{12.35}$$

$$V = \frac{V_{CC}}{3} - V_B \tag{12.36}$$

The R_2C_2 low pass filter gives the average value of this pulse train V_N and is given as

$$V_A = \frac{1}{T}\int_0^{T_1} V_1 dt$$

$$V_A = \frac{V_1}{T} T_1 \quad (12.37)$$

Equations (12.33)–(12.36) in (12.37) give

$$V_A = \frac{V_1^2}{V_B} \quad (12.38)$$

The output of the summer SU_2 will be

$$V_B = V_O + V_2 \quad (12.39)$$

The output of the summer SU_1 will be

$$V_O = V_A + V_2 \quad (12.40)$$

Equations (12.38) and (12.39) in (12.40) give

$$V_O = \frac{V_1^2}{V_O + V_2} + V_2 \quad (12.41)$$

$$V_O^2 = V_1^2 + V_2^2$$

$$V_O = \sqrt{V_1^2 + V_2^2} \quad (12.42)$$

The values of R_O and R_1 should be such that $R_O << R_1$.

12.5 TIME DIVISION VMC WITH NO REFERENCE—TYPE V

The VMC using the time division principle without using any reference clock is shown in Figure 12.9, and its associated waveforms are shown in Figure 12.10.

Initially the 555 timer output is HIGH. Pin 7 is opened. The multiplexer M_1 connects $-V_B$ to the low pass filter (LPF) ('ay' is connected 'a'). The output of the differential integrator will be

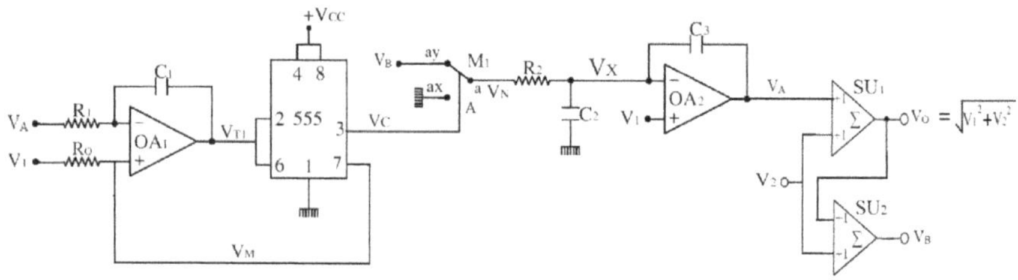

Figure 12.9 VMC without reference clock.

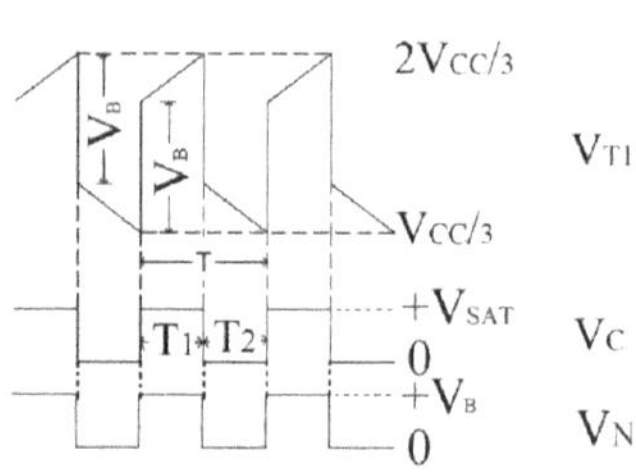

Figure 12.10 Associated waveforms of Figure 12.9.

$$V_{T1} = \frac{1}{R_1C_1}\int (V_1 - V_A)dt + V_1$$

$$V_{T1} = \frac{(V_1 - V_A)}{R_1C_1}t + V_1 \quad (12.43)$$

The output of the differential integrator rises toward positive saturation, and when it reaches the voltage level of $+2V_{CC}/3$, the 555 timer output becomes LOW. The multiplexer M_1 connects +0V to the low pass filter (LPF). Now the output of differential integrator will be

$$V_{T1} = \frac{1}{R_1C_1}\int (0 - V_A)dt + V_1$$

$$V_{T1} = -\frac{V_A}{R_1C_1}t + V_1 \quad (12.44)$$

The output of the differential integrator reverses toward negative saturation, and when it reaches the voltage level $V_{CC}/3$, the 555 timer output becomes HIGH, and the cycle therefore repeats, to give an asymmetrical rectangular wave V_C at the output of the 555 timer.

From the waveforms shown in Figure 12.10, it is observed that

$$T_1 = \frac{R_1C_1}{V_1 - V_A} V \tag{12.45}$$

$$T_2 = \frac{R_1C_1}{V_A} V \tag{12.46}$$

$$T = T_1 + T_2 = \frac{R_1C_1V_1V}{(V_1 - V_A)V_A} \tag{12.47}$$

$$V = \frac{V_{CC}}{3} - V_1 \tag{12.48}$$

Another rectangular wave V_N with $\pm V_B$ as the peak to peak value is generated at the multiplexer M_1 output. The R_2C_2 low pass filter gives the average value of this pulse train V_N and is given as

$$V_X = \frac{1}{T}\int_0^{T_1} V_B dt$$

$$V_X = \frac{V_B}{T} T_1 \tag{12.49}$$

Equations (12.45) and (12.47) in (12.49) give

$$V_X = \frac{V_A V_B}{V_1} \tag{12.50}$$

The op-amp OA_2 is at a negative closed loop configuration, and a positive dc voltage is ensured in the feedback loop. Hence its non-inverting terminal voltage is equal to its inverting terminal voltage, i.e.,

$$V_1 = V_X \tag{12.51}$$

From equations (12.50) and (12.51),

$$V_A = \frac{V_1^2}{V_B} \tag{12.52}$$

The output of the summer SU_2 will be

$$V_B = V_O + V_2 \tag{12.53}$$

The output of the summer SU_1 will be

$$V_O = V_A + V_2 \tag{12.54}$$

Equations (12.52) and (12.53) in (12.54) give

$$V_O = \frac{V_1^2}{V_O + V_2} + V_2 \tag{12.55}$$

$$V_O^2 = V_1^2 + V_2^2$$

$$V_O = \sqrt{V_1^2 + V_2^2} \tag{12.56}$$

12.6 TIME DIVISION VMC WITH NO REFERENCE—TYPE VI

The VMC using the time division principle without using any reference clock is shown in Figure 12.11, and its associated waveforms are shown in Figure 12.12.

Initially the 555 timer output is HIGH, pin 7 of the 555 timer is opened, and the control amplifier OA_2 will work as a non-inverting amplifier. Hence $-V_1$ is given to the differential integrator OA_1. The output of the differential integrator, composed of the resistor R_1, capacitor C_1, and op-amp OA_1 will be

$$V_{T1} = \frac{1}{R_1C_1}\int V_B + V_1)dt$$

$$V_{T1} = \frac{(V_1 + V_B)}{R_1C_1}t \tag{12.57}$$

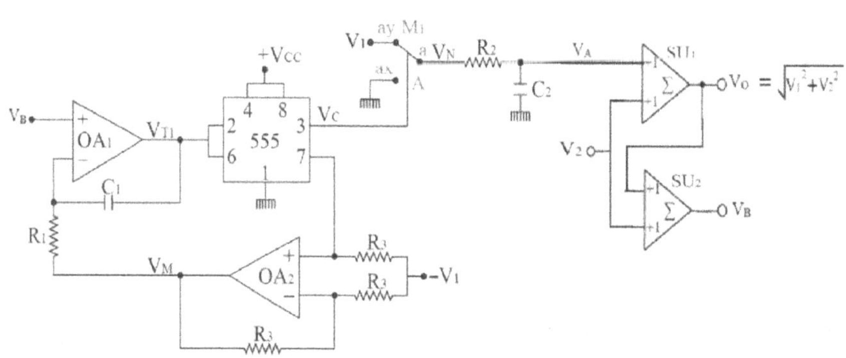

Figure 12.11 Time division VMC without reference clock.

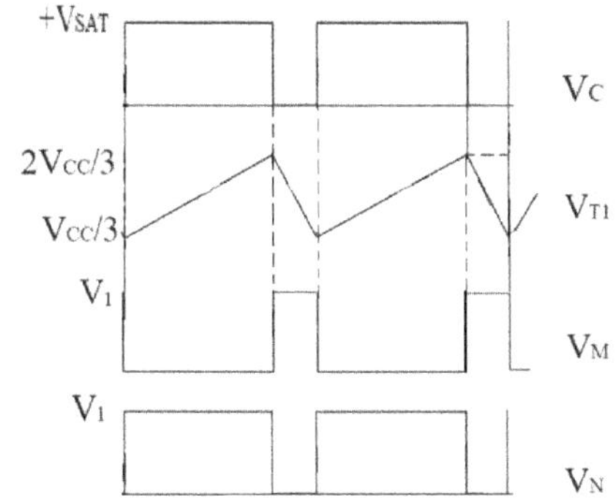

Figure 12.12 Associated waveforms of Figure 12.11.

The output of the differential integrator rises toward positive saturation, and when it reaches the voltage level of $2V_{CC}/3$, the 555 timer pin 3 output becomes LOW. Pin 7 of the 555 timer goes to GND. The control amplifier OA_2 will work as an inverting amplifier, and hence V_1 is given to the differential integrator OA_1. Now the output of the differential integrator, composed of the resistor R_1, capacitor C_1, and op-amp OA_1, will be

$$V_{T1} = \frac{1}{R_1C_1}\int (V_B - V_1)dt$$

$$V_{T1} = -\frac{(V_1 - V_B)}{R_1C_1}t \tag{12.58}$$

Let us assume $V_1 > V_B$.

The output of the differential integrator reverses toward negative saturation, and when it reaches the voltage level $(V_{CC}/3)$, the 555 timer output becomes HIGH, and the cycle therefore repeats, to give (1) an asymmetrical rectangular wave V_C at the output of the 555 timer and (2) another asymmetrical rectangular wave V_N at pin 7 of the 555 timer with the peak value of V_1.

The ON time of the rectangular wave V_N is given as

$$\delta_T = K\frac{V_1}{V_B}T \tag{12.59}$$

Another rectangular wave V_N is generated at the multiplexer M_1 output. The R_2C_2 low pass filter gives the average value of this pulse train V_N and is given as

$$V_A = \frac{1}{T}\int_0^{\delta_T} V_1 dt \quad (12.60)$$

$$V_A = \frac{V_1}{T}\delta_T \quad (12.61)$$

Equation (12.59) in (12.61) gives

$$V_A = \frac{V_1^2}{V_B}K \quad (12.62)$$

where K is a constant value.

The output of the summer SU_2 will be

$$V_B = V_O + V_2 \quad (12.63)$$

The output of the summer SU_1 will be

$$V_O = V_A + V_2 \quad (12.64)$$

Equations (12.62) and (12.63) in (12.64) give

$$V_O = \frac{V_1^2}{V_O + V_2} + V_2 \quad (12.65)$$

$$V_O^2 = V_1^2 + V_2^2$$

$$V_O = \sqrt{V_1^2 + V_2^2} \quad (12.66)$$

Index